SpringerBriefs in Philosophy

SpringerBriefs present concise summaries of cutting-edge research and practical applications across a wide spectrum of fields. Featuring compact volumes of 50 to 125 pages, the series covers a range of content from professional to academic. Typical topics might include:

- A timely report of state-of-the art analytical techniques
- A bridge between new research results, as published in journal articles, and a contextual literature review
- A snapshot of a hot or emerging topic
- An in-depth case study or clinical example
- A presentation of core concepts that students must understand in order to make independent contributions

SpringerBriefs in Philosophy cover a broad range of philosophical fields including: Philosophy of Science, Logic, Non-Western Thinking and Western Philosophy. We also consider biographies, full or partial, of key thinkers and pioneers.

SpringerBriefs are characterized by fast, global electronic dissemination, standard publishing contracts, standardized manuscript preparation and formatting guidelines, and expedited production schedules. Both solicited and unsolicited manuscripts are considered for publication in the SpringerBriefs in Philosophy series. Potential authors are warmly invited to complete and submit the Briefs Author Proposal form. All projects will be submitted to editorial review by external advisors.

SpringerBriefs are characterized by expedited production schedules with the aim for publication 8 to 12 weeks after acceptance and fast, global electronic dissemination through our online platform SpringerLink. The standard concise author contracts guarantee that

- an individual ISBN is assigned to each manuscript
- each manuscript is copyrighted in the name of the author
- the author retains the right to post the pre-publication version on his/her website or that of his/her institution.

More information about this series at http://www.springer.com/series/10082

Mario Graziano

Dual-Process Theories of Numerical Cognition

 Springer

Mario Graziano
Department of Cognitive Sciences
University of Messina
Messina, Italy

ISSN 2211-4548   ISSN 2211-4556   (electronic)
SpringerBriefs in Philosophy
ISBN 978-3-319-96796-7   ISBN 978-3-319-96797-4   (eBook)
https://doi.org/10.1007/978-3-319-96797-4

Library of Congress Control Number: 2018949060

This Springer imprint is published by the registered company Springer Nature Switzerland AG
The registered company address is: Gewerbestrasse 11, 6330 Cham, Switzerland

*For my two children, Salvatore and
Giuseppe; and my wife, Mimma.*

About This Book

As confirmed by a series of experimental data, there are two different cognitive systems relating to mathematical skills. The first system is not based on symbols, and it is approximative; it is based on the estimation of quantities; and it involves both a simple process of comparison and a series of basic arithmetical operations like addition and subtraction. The second system is based on symbols, and it is language- and culture-dependent; it is typical of adults; and it is founded on the ability of counting, therefore on a numerical system and on all arithmetical operations.

Therefore, to explain the acquisition of mathematical concepts, we must answer the two following questions. How can the concepts of approximate numerosity become an object of thought that is so accessible to our consciousness? How are these concepts refined and specified in such a way as to become numbers? Unfortunately, starting from these experimental results, there is currently no model that can truly demonstrate the role of language in the development of numerical skills starting from approximate pre-verbal skills. The aim of this book is to answer these difficult questions by turning to the dual process theories. This theoretical approach is widely used by theorists focusing on reasoning, decision making, social cognition, consciousness, etc. In this book, for the first time this theoretical approach is applied to the studies on mathematical knowledge with the aim of detailing the results brought about by psychological and neuroscientific studies conducted on numerical cognition by a few neuroscientists and laying the foundations of a new potential philosophical explanation on mathematical knowledge.

Contents

Part II   The Transition from System 1 to System 2

Chapter 1
Introduction

Using numbers to trade, classify, and sort goods and items may seem an easy, convenient, and almost ordinary task. This fact in itself could seem surprising, since numbers are—as Adam Smith famously said,—"among the most abstract ideas which the human mind is capable of forming". If it were to be true, then, using numbers would require a long and difficult training. Nevertheless, anyone can count and perform simple arithmetic calculations.

We are hence inclined to ask ourselves: what is a number? How are numbers represented in our mind? How can we perform more and less complex mental calculations?

This book will try to answer these questions from a cognitive point of view, meaning that it will not simply focus on the definitions of mathematical concepts and their axioms, but it will rather dwell upon how these concepts and axioms can be understood. In other words, it will try to account for the ideas and cognitive mechanisms underlying computation and the possibility of establishing axioms.

So, while the philosophers of mathematics try to define the meaning of theorems and establish how important axioms are, cognitive scientists look for the cognitive mechanisms that allow for the understanding of theorems or the production of mathematically true or false propositions.

The label "cognitive sciences" applies to a field of study where the multidisciplinary research activity on cognitive processes converges. Cognitive sciences embrace researchers coming from different disciplines, such as philosophy, neuroscience, psychology, evolutionary biology, linguistics, AI (only to mention those that contributed the most to the research in the field). The collection of disciplines involved is so broad and heterogeneous that it is necessary to talk about "cognitive sciences", emphasis on the plural form. This fact has its pros and cons: on the one hand, despite their different background, all researchers share a profound interest for the analysis of cognition; while on the other hand, according to their specific field of study, this analysis is carried out using methods that can strongly differ. This methodological pluralism is seen by many authors as a positive factor that fosters exchange, dialogue, and agreement, while others see it as a source of confusion and concern.

© The Author(s) 2018
M. Graziano, *Dual-Process Theories of Numerical Cognition*,
SpringerBriefs in Philosophy, https://doi.org/10.1007/978-3-319-96797-4_1

Another fact that raises some concerns is that some researchers might even argue that a research activity should not have the aim of studying the "cognitive structure of mathematics", since this endeavour runs the risk of becoming too vague or faced with too many difficulties (just think about the question "which mechanisms in the human brain and mind allow humans to conceive mathematical ideas and develop reasoning processes that follow mathematical criteria?") or, even worse, leading to false statements ("Is mathematics really based only on mind-brain mechanisms? What about Platonic mathematics? Does it not belong to mathematics?").

Yet, despite all these constraints, cognitive researchers are convinced that their work is of paramount importance, since the beauty and depth of mathematics can often seem inaccessible (particularly to non-mathematicians) because of the lack of a description of the cognitive structure of mathematics or because of the lack of a description of the brain and mind mechanisms that allow humans to conceive mathematical ideas and reason according to mathematical criteria.

Besides, beyond the epistemological problems of cognitive sciences (and its methodologies) and the doubts raised by foundationalist-platonic schools, the thesis of the embedded character of mathematics (in the broad meaning of mind used by cognitive sciences) is anything but new. As a matter of fact, going back to Aristotle and Greek arithmetic (and to the pebbles that gave birth to computation tasks), the whole empirical tradition focusses on a specific idea: mathematics starts with the questions and problems related to the combinatory and symbolic aspects of human experience. Starting from this acknowledgment, the argument develops into a deductive analysis of a huge number of formal structures that are strongly different, but mutually bonded.

Even a hard-line supporter of formalism such as the contemporary philosopher of mathematics Saunders MacLane claims that:

> we conclude that mathematics started from various human activities which suggest objects and operations (addition, multiplication, comparison of size) and thus lead to concepts (prime number, transformation) which are then embedded in formal axiomatic systems (Peano arithmetic, Euclidean geometry, the real number system, field theory, etc.). These systems codify deeper and non-obvious properties of the various originating human activities (Mac Lane 1981, p. 463).

Following the same rationale, Reuben Hersh (1997) tried to debunk the myths of mathematics (unit, universality, certainty, and objectivity), in order to replace them with ideas such as the human nature of mathematics, its fallibility, etc. According to Hersh, mathematical objects were not created arbitrarily by humans, but they were rather established starting from the activities that could be performed with pre-existing mathematical objects and the needs related to their current scientific activities and daily life.

So, starting from the agreed fact that mathematics is the product of the abilities of mankind (a point only the Platonists disagree with) and therefore its mental skills, it is fair to believe that the analysis should focus on this assumption and therefore on the cognitive mechanisms (possibly cerebral) underlying mathematical skills.

As it is well known, one of the earliest "modern" cognitive theories on the genesis of the concept of number was postulated by Jean Piaget (1952), who claimed that there is an unbreakable bond between the structure of general intelligence and the development of numerical competence.

Contrary to Piaget's claims, the cognitive studies carried out over the last decade of the 20th century showed that children have abilities of numerical cognition already at birth.

Despite the fact that children acquire the majority of their numerical knowledge through language, this does not seem to hinder the learning of natural numbers as it could be expected. The likeliest reason for this fact is that humans have two numerical representation systems at their disposal: one is "inborn-approximative", while the other is influenced by culture, "language-dependent" and at the basis of exact knowledge (Dehaene 1997). This assumption has been supported by abundant experimental data that highlight the existence of a sound numerical knowledge well ahead of the onset of the linguistic stage. Several tests showed that young children —so young as not to have any knowledge of numbers or computational methods as established by their cultures—are able to make mental representations of several numbers and have procedures that allow them to process these numerical representations in order to obtain more information (Xu and Carey 1996; Dehaene 1997). Children placed in front of advanced devices that offer the solution to simple problems of addition and subtraction react with surprise when the result seems "false" to them (Wynn 1990).

These inborn abilities are incredibly simple and do not include abstract mathematics, which can often even be associated to fear, but this does not mean that these skills are not as important as others. In inborn knowledge, a prominent role is played by the computational abilities related to the representation of the first three positive integer numbers, i.e. "one", "two", and "three". Dehaene claims that humans do not enumerate these numbers, but that they rather immediately perceive their presence because they relate to quantities that our brain perceives effortlessly and without resorting to enumeration. The technical term used for this process is *subitizing*, a term stemming from the Latin word *subitus*, which indicates a swift and precise recognition process related to the numerosity of sets composed of a maximum of 6 items.

Even today the debate on how subitizing works is still in progress. Yet, what is now clear is that children have the inborn ability—even before the emergence of a verbal enumeration procedure that is shown by the representation of a cardinal value for a set of objects—to make a representation of the transformation of a set (addition and subtraction) and to understand the relations between two numerosities. The experimental data collected through several experiments have indeed shown that young children (and some animals) have preverbal numerical skills that allow them to learn from the events taking place in their environment (the first part of this book will focus on these experiments). Nevertheless, the outcomes of several experiments do not prove that children always make good use of these skills, nor that they have an infallible numerical competence. On the contrary, the data highlight that children (as well as adults in particular circumstances) are more inclined to use perceptive indexes or heuristic techniques that are less precise, but much less energy-demanding from a cognitive perspective.

Unfortunately, what is still unclear is how children are able to use precise numerical representations starting from non-verbal approximative representations. In other words, children have indeed access to this wide range of representations of approximative numerical quantities, but they are not aware of it. Now, it seems clear that the concept of number must be accessible to consciousness, since the children that have acquired it are able to use numbers to distinguish the elements belonging to a set of objects.

Therefore, accounting for the learning process of mathematical concepts requires clear answers to two questions. Question 1: how can the concepts of approximative numerosity become an object of thought and become consciously accessible? Question 2: How can these concepts be enhanced and specified in order to become discrete numbers?

Up to now, despite the effort and work of several authors (summed up in Chap. 5), these questions remain unanswered and there is no model that can fully account for the development of exact numerical competences starting from preverbal approximative skills.

The ideas put forward by cognitivist researchers seem indeed more focussed on the concept of numerosity, a term that normally refers to the sense of number and in particular to the sense of the size of a set and not the size of numbers. The core of the issue is that, in order to acquire the concept of number, what is needed is a word or a symbol that refers to that concept. The complexity of mathematics is indeed due to the potential role that several cognitive functions can play in relation to computation, functions such as attention, symbolic representation, oral and written language, visual and spatial abilities, perception, motivation, will, and other abilities.

However, referring to all these mental functions at once implies claiming that the establishment, transmission, and development of mathematical knowledge should be explained through the study of the "mind". The description of the nature of our mind and the knowledge of the mechanisms underlying the cognitive processes that allow us to perceive spatial, temporal, numerical measurements and sizes, know the world and act in it, can help us understand what defines the *Sapiens* as a species and makes it structurally and functionally different to other animal species. Nevertheless, in order to do so, we need a representation of our mental life that is less idealised and abstract, and more realistic and in line with real human (and non human) abilities. From this point of view, more than four decades of research and experiments on the cognitive mechanisms hidden behind our daily thinking and acting have shown that human intelligence is not made up by a single system, but rather that it works through two peculiar agents: System 1 and System 2.

System 1 is intuitive, impulsive, automatic, unconscious, fast, and sustainable from an ecological and economical (in terms of mental energy) point of view. It allows us to seamlessly detect the fear on the face of another person. System 2 instead works consciously, deliberatively, slowly, reflectively, and is costly from an energy perspective. Nevertheless, this system allows us to multiply (even with some effort) 18 × 190. Each system works in its competence domain. When System 2 is passive, a "psychodrama" unfolds (a term coined by cognitive psychologist Daniel Kahneman), where what is false seems true, illusions take the upper hand, and mistakes become the rule rather than the exception.

Of course, not all study fields of mathematics deal with exact numbers and, therefore, with System 2. Beyond the linguistic and conceptual aspects of mathematics, there is also the inborn and biologically based form of mathematics that cognitive scientists so elegantly describe; a form of mathematics perfectly framed in the natural adaptation process related to the surrounding environment. This fact should not come as a surprise: humans are social, linguistic, and culturally determined animals, but they are still animals. On the one hand, their animal nature is due to biological factors while, on the other hand, their human nature depends on their culture. Humans are therefore the only species with a "second nature", the cultural one. It seems completely logical, but it can be easily forgotten.

References

Dehaene, S. (1997). *The number sense. How the mind creates mathematics*. New York, Cambridge (UK): Oxford University Press, Penguin Press.

Hersh, R. (1997). *What is mathematics, really?*. New York: Oxford University Press.

Mac Lane, S. (1981). Mathematical models: A sketch for the philosophy of mathematics. *The American Mathematical Monthly, 88*(7), 462–472.

Piaget, J. (1952). *The child's conception of number*. New York: Norton.

Wynn, K. (1990). Children's understanding of counting. *Cognition, 36,* 155–193.

Xu, F., & Carey, S. (1996). Infants' concept of numerical identity. *Cognitive Psychology, 30,* 111–153.

Part I
The Cognitive Science of Numbers

Chapter 2
The System 1

Abstract This chapter is devoted to the discovery of the core abilities underlying human numerical cognition. Neuroscientists hypothesises that human beings are born with a "number sense" that they share with other animals and that this instinct is the expression of the functioning of a "mental organ", a set of brain circuits that exist also in other species. According to neuroscientist Stanislas Dehaene, this "mental organ" works as an accumulator, namely a kind of approximate counting device that allows us to perceive, store, and compare numerical quantities.

Keywords Numerical cognition · Innate knowledge · Accumulator
Number sense

2.1 Numerical Cognition in Animals

The search for the biological foundations of human knowledge has always been one of the main issues addressed by comparative psychology, with the twofold ambition of identifying, on the one hand, the cognitive skills on which evolutionary adaptations are based, and on the other hand understanding how knowledge has evolved in animal species (especially primates) and mankind. From this point of view, there is at least one good reason to study the knowledge related to the estimation of "quantities" in animals and the role it plays in their daily lives.

The ability to quickly assess numerical quantities is indeed useful in food-searching behaviours, for example to estimate the energy content provided by a specific quantity of fruit hanging on a tree. Being able to estimate the number of branches available to build a nest represents an advantage in terms of time consumption, movement, and general efficiency. These skills may be beneficial also in social situations, for example in the case of fights or clashes among members of the same group (something that usually happens among primates and superior vertebrates), where a precise estimation of the number of opponents or allies can drastically affect the final outcome.

© The Author(s) 2018
M. Graziano, *Dual-Process Theories of Numerical Cognition*,
SpringerBriefs in Philosophy, https://doi.org/10.1007/978-3-319-96797-4_2

Another good reason to dwell upon numerical cognition in animals from a comparative point of view is the search for the limits of these skills in animals. This exercise may provide useful insights into the specificity of human numerical cognition and, above all, help identify the conditions that have led to its development in humans. At the same time, this may lead to the explanation of human numerical skills in evolutionary terms. As a matter of facts, humans master a system that allow them to process quantities and, because of their animal nature, it is sensible to believe that the animals closest to the human evolutionary chain are endowed at least with a rudimentary version of the same system.

Nevertheless, things are not as simple as they might seem.

Cognitive and cognitivist approaches to numerical knowledge have been developed only over the last two decades and even cognitive science still focusses on a multi-disciplinary approach to the matter, representing still a "collection of different disciplines" that change according to the authors involved. This limitation is reflected in particular in the supremacy of some disciplines, such as the studies related to the functioning of the nervous system at molecular and neuronal level or to the functioning of the human brain at an abstract and reasoning level (language, logic, philosophy). The importance given to these subjects seems to overshadow the role played by other disciplines (such as ethology) that are simply considered a natural enlargement to widen the research on the issues addressed by harder disciplines. Furthermore, by identifying cognitive science as a "collection of different disciplines" and labelling it as "cognitive sciences", two major problems arise.

Firstly, it leads to the consideration that cognition is an exclusive human skill, an idea that has been supported for years and clearly based on an anthropocentric approach. Secondly, it does not allow to bridge the old theoretical and methodological gap between neuroscience and neurobiology on the one side, and psychology on the other.

The success enjoyed by brain imaging technologies overshadowed the fact that even the most advanced imaging techniques—as well as other methodologies that aim at explaining the complex links between brain and behaviour—have in their theoretical toolbox a series of methods borrowed from psychology and have the goal of explaining observable data and curbing the possible multiplication of explanations of the phenomena considered.

This fact is of paramount importance, as behavioural models are applied to the interpretation of neurophysiological data. To overcome these issues, it is useful to make full use of the inquiries that aim at explaining identical phenomena at different levels of integration (e.g. from the simplest to the most complex), as well as presenting an evolutionary paradigm. As a matter of fact, the "great chain of being" (*Scala naturae*) is based on a presumed complexity of the structure of living creatures, especially at the level of the nervous system. Humans are placed at the top of this chain because of the complexity of their brain.

Nevertheless, despite standing at the top of the chain, even humans are a mere evolutionary "product", questioning once again the assumption that cognition is an exclusive human skill. As other knowledge-building processes, also numerical representation has been biologically developed through selection pressures that, in the framework of an evolutionary process, have rewarded the individuals that could make the most advantageous choices in settings where their mathematical skills allowed them to solve specific problems.

This book will present naturalistic observations that emphasize the ecological importance of mastering calculation skills, as well as laboratory studies that, on the contrary, systematically investigate the limits and potential of these skills. In particular, we will focus on studies that used rigid training techniques, as well as those based on the observation of spontaneous preferences in an environment with controlled numerical variables.

2.1.1 From Anecdotes to Early Experiments

A number of tales describe the skills of amazing animals that could perform real mathematical wonders. One of them dates back to the 18th century and recounts the abilities of a crow that could count up to five, and of a farmer that decided to kill him to defend his crop from the dangerous ravishes of the animal. Nevertheless, every time that the farmer tried to approach the crow's nest built on top of a tower, the crow flew away, returning only when the farmer had left. To catch the animal, the farmer asked the help of his neighbour and they decided to ambush the crow in the tower together. After a while, one of them left the tower, but the animal did not fell into their trap and came back to his nest only after he saw that the second farmer had left the building. The farmer, then, asked two other men to join them, and then three. They kept failing, though, because before returning to his nest, the crow always waited for the last man to leave the tower. Finally, when six men got involved, the farmer won his battle: the animal waited for the exit of five men and then confidently returned to his nest where he got killed by the sixth man, who was waiting for him.

Nevertheless, the most famous example of this kind of tales is the story of Hans, a male Arabian horse of Russian lineage that was bought for few pennies by his master, baron Wilhem von Osten, because of a small physical defect. At the beginning of the 20th century in Berlin, baron von Osten stated that, after ten years of hard training, he had taught arithmetic to his horse. The news spread all over Germany and both Hans and his master became real stars. But what was Hans capable of? When Von Osten wrote two on a blackboard, Hans tapped his hoof twice on the ground. When the number was three, Hans did it for three times and so on, up to the number of ten.

Encouraged by Hans' achievements, his master tried to teach him more difficult assignments, such as additions and subtractions, square roots, and finally fractions. Unfortunately for von Osten, the sceptical German academics ordered an inquiry on the issue, despite representing a minority compared to the majority of their easy-to-fool fellow citizens. The *Hans Commission* was established with the involvement of two zoologists, the psychologist Carl Stumpf, and a famous horse tamer, who were to establish whether something had been fixed. After several tests, in 1904, the commission came to the conclusion that there were no tricks and that the intelligence and skills of the horse were genuine. Nevertheless, this conclusion did not satisfy Oskar Pfungst, a pupil of Stumpf, who insistently demanded Hans to be subject to a new series of tests. Pfungst focussed in particular on two cases. In the first one, Pfungst asked the person that formulated the arithmetical question to Hans to step back from the horse, noticing immediately a drastic reduction in the number of correct answers given by the animal. In the second case, he asked that the person formulating the question should ignore the right answer to his own question. The result was that the right answers given by the horse went down to virtually zero. Pfungst's test highlighted the fact that Hans did not master mathematical concepts, but rather that the horse, whenever a question was asked, had an incredible ability to "detect" the breath, posture, and facial expression of the person asking the question and knowing the right answer. When Hans started tapping his hoof on the ground to provide an answer, the slight tension appearing on the face of the interviewer allowed Hans to stop whenever the answer was satisfactory.

Hans' case emphasized the importance of fine-tuning the psychological experiments on animals coupled with the need of developing much more rigorous research methods. The first convincing examples of animals that owned genuine numerical skills date back to the 1950s, with the studies conducted by a psychologist called Francis Mechner. In Mechner's experimental setting, mice were deprived of food for a short period of time before entering a cage containing two levers (A and B). In order to get the food, the mouse had to pull lever A for a given number of times (n) before pulling lever B. If the mouse pulled lever B before pulling A the right n number of times, it would not receive any food and it would also receive a slight electrical shock. By changing n, Mechner wanted to prove that mice were able to change their behaviour and, as a matter of fact, the average number of times that the mice pulled lever A was strongly correlated to the required minimum number (n). In other words, if the experiment required the mice to pull lever A four times before pulling lever B, over time the mice learnt to pull lever A "more or less" four times before pulling B. It is important to note that the animals never learnt to pull A exactly four times before B, but rather they tended to overestimate n by pulling the lever four, five, or six times. Similarly, the mice involved in an experiment requiring them to pull lever A eight times before B, learnt to pull it almost eight times (Mechner 1958). A decade later, Platt and Johnson (1971) achieved the same results on numbers (n) between 4 and 24. Nevertheless, despite the coherence of the outcomes of the two studies, a general remark ought to be made: how can we be sure that the mice "answered" on the basis of the number of times they pulled the lever and not because of other factors, such as the time spent pulling the lever or the total energy consumed?

A couple of years after the publication of his first study, Mechner and his colleague Laurence Guevrekian tried to answer this difficult question by developing a series of experiments in which mice were exposed to a situation of water deprivation. Mechner proved that the thirstier the mice were, the faster they pulled the levers. Therefore, the time interval between the first pulling of lever A and the pulling of lever B decreased as the mice got thirstier. Nevertheless, despite the faster pulling pace, the mice that had learnt to pull A four times kept doing so. This fact led the researcher to the conclusion that time was not affecting their actions: the mice were making an estimation of the right number. Similarly, in a control test where mice had to wait a specific amount of time to pull B after pulling A, water deprivation did not affect the final results (Mechner and Guevrekian 1962).

Starting from these early experiments, other authors tried to carry out new studies where different species were tested from a visual, auditory, tactile, and kinaesthetic perspective, with the aim of understanding whether the numerical performances of animals were due to an abstract conception of numerosity or to other perception skills that depended, in some way or another, on the presentation modality. Following this rationale, Fernandes and Church (1982) developed an interesting experiment on mice that required them to distinguish sequences of 2 or 4 sounds. Once again, the mice had to enter a cage with two levers, one related to "2" and the other to "4". To be sure that the animal chose one lever only on the basis of numerical parameters, the researchers placed great care in deciding the stimuli associated to each number (2 and 4), systematically assessing the duration of each sound, the time interval between sounds, the rhythm of the sequence, and the total sound energy. After two years, Church and his colleague Meck decided to address the same question but with a different methodology. They trained a group of mice to pull a lever when two sounds were heard and another lever in the case of eight sounds (Meck and Church 1984). In the first stage of this experiment, Meck and Church exposed the mice exclusively to stimuli that had a perfectly correlated relationship as regards number and duration: the first sequence had two sounds and lasted for 2 s, while the second one had eight sounds and lasted for 8 s. The mice therefore learned how to associate these two sequences to two different levers. They were then exposed to new sequences: in the first session, the number of the sequence was the same (4) but the duration varied between 2 and 8 s. In another session, instead, the duration of the sequence was the same (4 s), while the number of sounds changed between two and eight. The results proved that mice were able to generalise the associations learnt during the preliminary stage of the experiment both as regards the duration of the sequence and the number of sounds.

In a different experiment, the sounds were replaced by light flashes: under these new experimental conditions, the mice had to pull a first lever when exposed to two flashes and a second lever when exposed to four flashes. Despite needing much more time in the preliminary stage of the experiment to establish the association between these actions compared to the previous experiment, the animals performed correctly, proving that they had a representation of numbers that did not depend on a specific presentation modality.

Encouraged by these results, the researchers performed a test of number recognition on mice by switching between visual, auditory, and tactile presentation modalities (Fig. 2.1).

In this case, the mice initially underwent a recognition test of sequences of 2 or 4 sounds. Then, they were immediately tested by using visual stimuli (sequences of flash lights). By comparing these mice with those belonging to a control group, the researchers noticed that, in the first group of mice, the lever associated to two

Fig. 2.1 Mice's answers in the experiment developed by Meck and Church (Meck and Church 1984)

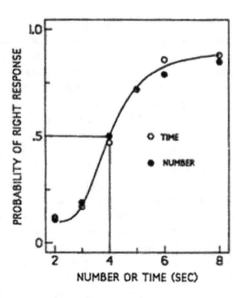

sounds was also associated to 2 flashes, while in the control group the two levers were often mixed up. Meck and Church believed that the difference observed in the two groups depended on the fact that the first group of mice successfully applied the distinction learnt through auditory stimuli to the visual stimuli. Later on they proved the same also by exchanging auditory stimuli with tactile ones, achieving the same outcomes (in this test, the stimuli containing 4 sounds were replaced by 4 electric shocks).

Some researchers have however questioned the outcomes of these experiments. For example, Davis and Albert (1987) performed a test on a group of mice to understand if they could differentiate between sequences composed of 3 following sounds and 2 or 4 sounds. The same mice were then exposed to stimuli of 2, 3, or 4 flashes. In this experiment, the authors did not notice a transfer from visual to auditory stimuli of the ability related to the distinction of numerosity. Davis and Perusse (1988) came to the same conclusion, stating that, despite the fact that animals had learnt to distinguish different numerosities upon hard training, they did so only as a last-resort solution, i.e. when they were deprived of all other kinds of information. For example, when a mouse had to perform an action and had the possibility to decide whether to distinguish between numerosities or, on the contrary, assess the area of a specific surface, the animal decided to assess the area of the surface, thus ignoring numerosity. Furthermore, as far as the experiment by Meck and Church is concerned, the researchers put forward the idea that mice took their decisions according to the quantity of energy contained by the stimuli. In this way, the stimuli containing two sounds (that had less sound energy compared to the stimuli containing 4 sounds) were associated to the stimuli of 2 flashes, which contained less light energy compared to the stimuli of 4 flashes.

For all these reasons, we may come to the conclusion that the results of these early experiments are questionable: some of them show that animals can transfer their numerical skills from one modality to the other, while others show on the contrary that such transfer does not happen. What is clear, though, is that the tasks assigned to animals require a certain amount of training to be mastered. Before succeeding, the animals have to undergo hundreds of training sessions. It is therefore to assume that, despite animals can distinguish between different numerosities when forced to do so, their natural behaviour is completely different. Research papers have shown that numerical discrimination tasks based on spontaneous preferences record worse performances by animals compared to those made following strict training procedures (Hauser and Spelke 2004). This fact could mean that numerosity is not a feature that animals use as a natural element. Nevertheless, this hypothesis stands against the evolutionary arguments that contend that the representation of numerosity is a selective trait existing in animals in a more or less rudimentary form. Several studies have highlighted behaviours in the wild where quantification mechanisms seem to be in place, for example when animals search for food or fight.

A good example of this is the experiment performed by Karen McComb and colleagues (McComb et al. 1994), which showed that lions living in the wild—in the Serengeti National Park in Tanzania—moved aggressively towards a place where food was available only when they knew that they were more numerous than their possible enemies, whose roars were played by using taped recordings, while they avoided moving there in the case their pride was the smallest one. This experiment showed that lions have the ability of comparing the number of roars heard with the number of lions belonging to their pride. Therefore, lions seemed able to make abstract representations of numbers, independently of the presentation modality or the features of the stimuli to which they are exposed. The same was recorded also in male chimpanzees, who normally attack other neighbouring groups only if their cartload is bigger enough to provide muscle power to their attack (Wilson et al. 2001).

Unfortunately, the tests of spontaneous choice must deal with an important theoretical limit that hinders the comparison with the studies based on training procedures: if a choice between two numerical stimuli is an index of the capacity to distinguish numerosities, the lack of a choice does not directly imply the lack of this skill. For example, an animal may decide that it is better to pick a group of 4 oranges compared to a group of 3, but at the same time the same animal may decide to go for a group of 9 oranges instead of a group of 10, despite owning all cognitive structures needed to distinguish 9 from 10. Despite all these facts, the observation of spontaneous choices has advantages compared to laboratory training procedures, such as the possibility to postulate hypotheses on the natural environment where these skills are used, leading to assumptions on the evolutionary importance of having a number sense and on the animals' actual use of these skills in their natural environment.

2.1.2   Birds and Mammals

Several studies have tried to emphasize how birds may use the concept of number in their daily lives. An example is represented by coots that, according to a specific study, use the concept of number when deciding whether they should lay one more egg or not, establishing a comparison between the number of laid eggs and the number of eggs belonging to other brood parasite birds (Lyon 2003). The study was made by observing the coot in its natural habitat and it showed that these birds are able to use a calculation strategy to take out the eggs belonging to other birds from their nests, while at the same time keeping the number of eggs necessary to maximize their *fitness*. One of the earliest researchers that noticed the numerical skills of birds was German zoologist Otto Koehler. In mid-20th century, he proved that birds master excellent skills in comparing the size of two groups presented to them at the same time, as well as in remembering the number of the objects belonging to a sequence.

In one of his experiments, Koehler trained a crow to recognize the number of dots present on a piece of cardboard, which was the same as the number placed on a box containing food. The researcher proved that crows could make a distinction between 2, 3, 4, 5, and 6 dots. In another experiment, in order to get some food, jackdaws had to open boxes that contained up to four or five units of food. The boxes where placed randomly to avoid animals using non-numerical parameters to make their choice (such as the length of the row of boxes that had to be opened) and each and every one of them contained 0 to 2 units of food. In this way, animals had to base their choice on the number of units they had already taken (Devlin 2005).

Following a similar methodology, Emmerton and Delius (1993) tested some pigeons to evaluate their ability to distinguish groups of dots that differed only by one unit, such as 1 *vs.* 2, 2 *vs.* 3, 3 *vs.* 4, and so on. The results of this experiment showed that they could tell them apart up to the group 6 *vs.* 7. A few years later, Emmertong and some colleagues decided to make another experiment with a different modality (Emmerton et al. 1997). In a preliminary training stage, pigeons learned how to distinguish two different types of stimuli and associate them to a lever. The stimuli containing one or two elements were associated to lever A, while those containing 6 or 7 were associated to lever B. During the test, pigeons were exposed to groups composed of 3, 4, or 5 elements (groups of dots) and the birds had to react using the same levers. Both experiments emphasized that pigeons could apply the discrimination they had learnt also to stimuli that they had never encountered.

One more evidence of the numerical skills possessed by birds was provided by the research by Irene Pepperberg (2006), who trained a parrot to repeat the number of objects showed to him, a task that does not only require the ability to distinguish and discriminate between numerosities, but also to associate each number to a vocal response. Furthermore, many bird species display their numerical skills when counting the repetitions of each note in their songs. The songs of birds contain indeed stereotyped sound sequences and the dialectal varieties of these songs differ in the number of times that notes are repeated (Marler and Tamura 1962). The correct number of repetitions of notes is learnt and passed on to the next generation through continuous exposure. Therefore, even though several aspects of the song typical to a bird may be genetically determined, it seems that birds can estimate the length of each sound that they produce. A similar concept of understanding of ordinal numbers in birds was found also in chicks, in an experiment divided into four stages and performed by Rosa Rugani and colleagues (Rugani et al. 2007). In the first stage of the experiment, chicks were trained to peck the third, fourth, and sixth group of stimuli in a series of ten identical groups. In the following stages, the order of the groups changed, to avoid chicks using spatial parameters in their decisions. Despite this change, chicks successfully identified the position of the right group, proving that they did not rely on spatial indicators, but rather on an internal ordinal order that allowed them to identify the disk containing food among all possible alternatives. Regarding the numerical skills in mammals, several studies focussed on the discrimination skills in dolphins.

For example, Annette Kilian and colleagues (Kilian et al. 2003) performed a research inquiring the numerical cognition of water mammals, asking a *Tursiops truncates* dolphin to distinguish between two groups composed of respectively five and two items (Fig. 2.2).

Once again, to be sure that objects were distinguished solely on the basis of numerical parameters and not because of their shape or the generic disposition of the group, these items had different shapes and their position kept changing. The conclusion was that dolphins could identify the most numerous group even when faced with configurations that they had never met in training, showing that their choices were exclusively based on the assessment of numerical elements. A couple of years later, the same research group decided to investigate the possibility that dolphins had specialisation mechanisms at a hemispherical level that played a role in the processing of numerical information. In this experiment, two groups of stimuli with different numerosities were placed in the dolphins' tank, controlling the

Fig. 2.2 Representation of the experimental setting used to train dolphins. On the right, a detail of the stimuli used (Kilian et al. 2003)

shape and position of these two groups. In the first test, dolphins had to distinguish groups of 2 vs. 5 objects and the results proved that the animals could solve the task when the numerical stimulus was analysed through the right eye. In the following test, non-numerical parameters such as the shape and position of the stimuli were not controlled, and the researchers evaluated the performances of animals in distinguishing two numerical groups: 2 vs. 5 and 3 vs. 4. The results showed that dolphins could distinguish 2 vs. 5 both with the right and left eye, but in the case of 3 vs. 4 dolphins recorded better performances when the stimuli were analysed with the right eye. These data led the researchers to the conclusion that, because of the anatomical features of the optic tracts of dolphins, the left hemisphere of water mammals has specialised in distinguishing numerosities (Kilian et al. 2005).

Furthermore, it was proved that dolphins can make assessments based on the ordinal features of numbers. A group composed of Kelly Jaakkola and colleagues (Jaakkola et al. 2005) trained some dolphins using food rewards in order to make them choose the group containing the lowest quantity of stimuli between two options. Once learnt, the same rule was successfully applied to new configurations of stimuli that presented numerosities never experienced during training, hence showing that dolphins have cognitive structures suited to the recognition and representation of numerosities on an ordinal scale.

Despite a majority of studies investigating the existence of numerical skills in mammals and birds, there is also a study carried out by Claudia Uller and colleagues (Uller et al. 2003) that tried to find numerical discrimination abilities in red-backed salamanders (*Plethodon cinereus*), a surprising goal taking into account the evolutionary gap existing between this species and the others. In their experiment, the researchers placed the salamanders in a corridor with two boxes containing different quantities of food rewards, i.e. fruit flies, at the opposite ends of the corridor. The authors noticed that the salamanders always went towards the box containing the higher number of fruit flies, both in the cases 3 vs. 2 and 2 vs. 1.

On the contrary, this discrimination ability was not recorded in the 3 vs. 4 and 4 vs. 6 cases. Nevertheless, it is possible to question this study. As a matter of fact, when living creatures are used as stimuli (a common practice in ethological studies that use conspecific stimuli or potential preys), the amount of movement recorded in the stimulus-group can provide a direct indication of its size, hence helping in deciding which group is bigger. This is even more the case for salamanders, since their visual system reacts almost exclusively to movement (Robins et al. 1998). In the case of Uller's study, it is likely that one or more flies were moving and therefore the probability that movement could be perceived by the salamander in the box with three flies was bigger than in the group with two flies. These studies require procedures that can control perception factors. Only this kind of control would make it possible to make a general distinction between studies based on a generic discrimination ability of quantities and studies useful to highlight specific numerical discrimination abilities, in other words studies that might provide evidence of a real numerical calculation system in animals and do not leave any doubts on the fact that subjects could use other variables. Furthermore, according to some authors, it is not justified to talk about numerical cognition if the numerical representation involved cannot be added, subtracted, or at least ordered (Gallistel and Gelman 2000). Currently, the majority of studies dealing with this issue are almost exclusively those on the numerical skills of non-human primates.

2.2 Numerical Cognition in Primates

The most extensive research on the numerical skills of animals deals with primates and, taking into account their affinity to species such as humans, their performances have led to several comparisons to the behaviours seen both in children and adults. Stemming from the idea outlined by Gallistel and Gelman (2000), a fundamental experiment that aimed at evaluating whether chimpanzees could put numerosities in the right order was the one performed by Brannon and Terrace (1998). In their experiment, two *rhesus* monkeys were assessed according to their ability to put pairs of numbers in the right order when they were displayed on a monitor. These number pairs contained numbers from 1 to 9 and they were presented to them only after undergoing training sessions that taught them to put configurations containing 1 to 4 items in the right order. In order to ensure that animals reacted to numerosities and not to other non-numerical parameters, the researchers used a wide range of stimuli, trying to minimise non-numerical parameters. Furthermore, to establish a genuine test that could assess numerical ordinal skills, the monkeys were not rewarded in tests containing new numerical values, while they received a reward during the initial experimental tests containing two numerosities between 1 and 4. Initially, the monkeys learnt to touch groups containing from 1 to 4 units displayed on a monitor, going from the smaller to the bigger one. In a second stage, they were showed new groups (never displayed in the training stage), which contained 5 to 9 items. The results emphasized that monkeys, by generalising the concepts learnt during the previous experiments and applying them to the new situations they faced, were able to put the new images in the right order, touching at first the groups containing fewer elements and then those containing more, providing a first proof of their ability to make a representation of the numerosity of stimuli and the possibility to put them in a specific order. Nevertheless, even Brannon and Terrace (2000) noticed that, when asked to put the images in the opposite order (from the bigger to the smaller one), the animals were not able to generalise the concepts learned and successfully perform the task.

Similarly, the monkeys encountered the same difficulties when they were asked to put the images in a non-monotonous arbitrary order (for example, 4−1−3−2). Even after going through long and hard training sessions, they never learnt to react to an arbitrary order, despite the fact that they had no problem in putting the images in the right increasing order. To explain this fact, the authors put forward two possible explanations: either the increasing order is a concept so powerful in animals that it cannot be inhibited, or a non-monotonous order is too difficult because it requires the ability of identifying each single number before providing a reply. As a matter of fact, putting images in the right increasing order does not necessarily require the knowledge that a group contains 1 unit, 2 units, and so on.

The ability to recognise which set contains fewer elements than all the other groups, etc. is all it is needed. On the contrary, to put images in the right position in a 4−1−3−2 order, animals must recognise the cardinal features of each element. It is possible to overcome this issue by presenting monkeys with sequences of three numbers: in this way, the cardinality of numbers would not represent a problem and the animals could simply define each number as "the biggest", "the smallest", or "the number that is not the biggest nor the smallest". Unfortunately, the authors did not perform this kind of test on their monkeys.

Subsequently, Jordan and Brannon (2006) trained some *rhesus* macaques to "point" on a monitor the set of objects that was most similar to the numerosity of a reference sample showed to the animals at the beginning of the experiment. By recording reaction times and success rates, it was proved that macaques could solve this task, despite the fact that the variables time/accuracy are strongly affected by the "proportionality" law. According to this law, the level of confusion between two numbers (or two numerosities) recorded by the percentage of right answers in comparison tasks depends exclusively on the quotient of these two numbers. The closest to 1 the ratio between the numbers is, the likelier the two numbers will be confused by the subject. It is worth noticing, though, that this observation must not be valid when the formats of numbers are not controlled. For example, in the two pairs of numbers (10, 15) and (30, 40), the distance between the two numbers is bigger in the second group (40 − 30 > 15 − 10), but at the same time the quotient between the two numbers tends more towards 1 in the first group (1 < 40/30 = 4/3 < 15/10 = 3/2). The proportionality law therefore states the importance of format: once the distance between two numbers is established, their quotient tends more to 1, as the two numbers grow bigger.

The fact that proportionality law is often called Weber's law (or law of Weberian behaviour) without providing reasons for doing so has led many to mix up the two concepts. According to Weber's law, the distance separating one stimulus from the following distinguishable one (with a success rate of 75%) is proportional to the value of the reference stimulus. The concept can be expressed in the following mathematical terms: in a pair of stimuli with numerosity (N, N*a), the proportionality law states that the distinguishing rate of these values does not depend on N, but only on "a", and it may be expressed by the formula:

$$p\ (N,\ N^*a)\ =\ p\ (a)$$

On the contrary, according to Weber's law, the starting value is not (a), but rather an established performance threshold (75% of correct answers) that should be reached. Therefore, for a stimulus N, the performance level $p0$ is reached when it is possible to distinguish one stimulus from the other. In the stimulus NW, the value of W depends on the performance level ($p0$). If we postulate a level of ($p0$) = W and that (a) > (N, Na) may be prolonged in a continuous and decreasing function of (a), the two laws are equivalent. All numerosity assessments are therefore based on two principles that seem to play an important role both in animals and (as we will explore in the next chapter) in children and adults: the principle of "distance effect" and the "size effect" (also known as "magnitude").

Following the first principle, two numerical quantities are easier to distinguish when the distance between them increases. Therefore, it is easier to distinguish a group of two elements from a group of five, compared to a group of four elements from a group of five. The second principle, the "size effect", instead states that, when the distance between two numbers is the same, it is more difficult to distinguish them as they grow. Therefore, it is more difficult to determine which group is bigger when looking to two groups containing ten and eleven elements compared to two and three elements, despite the fact that the distance between the first and the second group is always equal to one.

As far as the mathematical addition skills of monkeys are concerned, one of the earliest experiments in this field was performed by primatologists Guy Woodruff and David Premack of the University of Pennsylvania, who became famous in 1978 with an article bearing the provocative title *Does the chimpanzee have a theory of mind?* that aimed at showing that chimpanzees could solve different problems by inferring scopes and intentions, laying the foundations of what was subsequently called the "Theory of Mind".

In one of their experiments that focussed on the possibility that chimpanzees mastered numerical skills, Woodruff and Premack (1981) asked chimpanzees to choose (by offering them food rewards) the item that was physically more similar to a third object, from a group of two objects. After this preparatory stage, the subjects were shown a glass half-filled with a blue liquid and asked to decide between two options: half apple or three quarters of an apple. The chimpanzees mostly chose the half apple, basing their decision on the conceptual similarity between the half-filled glass and the half apple, showing that they understood the concept of numerical fractions. To prove that their decision was based on the fraction that represented a numerical quantity and not on the volume of the glass filled with coloured liquid, the researchers asked the chimpanzees to perform an even more abstract task, showing that these animals could actually mentally combine two fractions.

In this second experiment, the sample stimulus was represented by a quarter of an apple and a half-filled glass, while the choice was between a full disk and three-quarters of a disk. In this case, the animals mainly chose the latter option, thus correctly solving the mathematical operation $\frac{1}{4} + \frac{1}{2} = \frac{3}{4}$. Later studies came to similar conclusions. An example is the experiment performed by Duane Rumbaugh and colleagues (Rumbaugh et al. 1987), where apes were put in front of four containers grouped in two and filled with chocolates. When the apes had to choose between the two groups of containers, in the majority of cases, they picked the group where the sum of the chocolates contained in the containers gave the highest total number of chocolates. The positive performances increased when the distance between the total number of chocolates in the two groups was bigger: the more the sum differed, the better the apes succeeded in their task. These results have been confirmed by an experiment performed by Sulkowski and Hauser (2001), who wanted to investigate the abilities of monkeys and apes in subtracting quantities of food. The authors used *rhesus* macaques raised in captivity to perform only one test, in order to avoid the possibility that the training stage for this test could lead the animals to the development of knowledge that they could use in the following stages of the experiment. In the test, the monkeys observed the researchers take out 0 or 1 plum from a group of 1−3 plums contained in a box. Then, the researchers left the room, leaving the monkeys to decide which box to pick. After 11 different tests, the authors noticed that the monkeys always picked the box containing the highest number of plums, even if this choice required them to pick the box that originally contained the lowest number of plums.

A similar methodology to that used by Sulkowski and Hauser was subsequently employed to evaluate the skills of chimpanzees in spontaneous mathematical additions (Beran and Beran 2004). Also in this case, the animals observed a researcher place one banana after the other into two identical boxes, without the ape seeing the total quantity of fruit in the boxes. In this way, if chimpanzees wanted to pick the box with the highest number of bananas, they had to perform a process of mathematical addition on the elements that they saw were placed in the box. Thanks to this simple procedure, it was proved that chimpanzees are able to perform spontaneous additions with small numerosities (1 banana *vs.* 2, 2 *vs.* 3, and 3 *vs.* 4) and bigger numerosities, in the latter case provided that the distance between the two groups is big enough (e.g. 5 bananas vs. 10, or 6 *vs.* 10).

In a more study by Hanus and Call (2007) that involved all four species of great apes, it was proved that they are able to pick the most numerous group both in the case of simultaneous spontaneous choice tests (when two groups are presented at the same time) and in tests where stimuli are presented successively. Furthermore, it was shown that the best index of the ability to distinguish groups was, one again, the ratio between quantities: as the ratio between two sets grew higher and hence the distance between the sets got smaller, the level of positive performance decreased. Contrary to this methodology, many studies on the spontaneous numerical skills of primates often use a research paradigm based on the cognitive studies performed on pre-verbal aged children, using the so called principle of "unattended expectation" (which we will address in the following chapter) that is de facto founded on the gap existing between the knowledge of test subjects and the reality they face.

For example, Jonathan Flombaum and colleagues (Flombaum et al. 2005) used this methodology to investigate the spontaneous calculation abilities in *rhesus* macaques. The researchers presented the animals with a box. They then proceeded to put some lemons into the box, adding them one by one. During the test, the entire content of the box was hidden behind a screen that was lowered only at a later stage, showing the result of the expected addition (or another incorrect result). In this test, the apes (not trained) stared more intensively at the box when the results differed "sufficiently" from the expected sum (8 and 4 are substantially different for apes, while the same is not true for 4 and 6). When facing a situation where the principles of numerical addition were violated, the staring time was longer compared to that of mathematically correct situations. Similar results were achieved with another primate, the tamarin (Uller et al. 2001), as well as four different species of prosimians (Santos et al. 2005). Subsequently, Cantlon and Brannon (2006) used the same experimental paradigm altering the presentation of the stimuli presented, which took place through a computer animation. In this case, animals were placed in front of a screen where images containing groups of points were shown. After a short break, the animals were shown a new group of elements and then, finally, two other images were shown: one containing a numerosity that was equal to the addition presented and the other containing a wrong answer. Each session was made up of 250 tests and several sums were presented (1 + 1 = 2, 4 or 8; 2 + 2 = 2, 4 or 8; 4 + 4 = 2, 4 or 8, up to all possible combinations of summands of the numbers 2, 4, 8, 12 and 16). To be sure that the animals picked the right answer only on the basis of numerical parameters and not other factors (density of the stimulus, surface of the points), the researchers presented the same sum by often changing the density of the stimulus. Furthermore, it was noticed that monkeys did not fail at their task even when the surface of points (contrary to numerical sums) was equal to the wrong sum. The only factor influencing the level of performance was the format of numbers presented: the level of performance concerning the addition 1 + 1 was statistically better than the one recorded for 2 + 2.

From several perspectives, therefore, the numerical skills of monkeys and apes resemble the abilities of numerical assessment recorded in humans. The main difference is of course the fact that humans can count with precision by using symbols that identify numbers. Starting from this consideration, some authors decided to embark on the ambitious project of teaching these symbols to chimpanzees. One of the earliest experiments in this sense was performed by Biro and Matsuzawa (2001), who trained a chimpanzee to associate Arabic numerals to a series of points displayed on a monitor. The experiment was divided into three stages. In the first stage, a group of dots and two Arabic numerals were displayed on the monitor, asking the animal to touch the Arabic numeral that identified correctly the quantity of dots displayed. In the second stage, which reflected the same concepts of the previous one, an Arabic numeral was displayed, followed by two groups of dots. In this case, the animal had to touch the group of dots that corresponded to the Arabic numeral. Finally, in the last stage, two Arabic numerals were displayed and the animal had to touch the two stimuli starting from the smaller one. The outcomes of this experiment showed that chimpanzees could solve mathematical tasks that required both a notion of cardinal numbers (the ability to associate the numerical symbol 3 to the category of elements involving three units) and ordinal features (the ability to identify the biggest element). Several studies reached the same conclusions. Among them, there is the extraordinary research made by Boysen and Berntson (1989) on a female chimpanzee called Sheba. The researchers trained Sheba to associate numerosities with Arabic numerals. In the first part of the experiment, they presented Sheba with oranges on a non-simultaneous basis (i.e. the oranges were not visible at the same time). Sheba's task was to indicate how many oranges she saw, pointing to the corresponding number. In the second stage of the experiment, Sheba did not find oranges in the boxes, but rather Arabic numerals. Despite this fact, she was able to provide the right answer to the addition required by the test. In the second version of this experiment the researchers provided Sheba with a bunch of cards, on which the numbers from 1 to 9 were printed. In this case, Sheba had to associate each card to a group of items containing 1 to 9 objects. Also in this case, the animal succeeded in her task and, in addition to this, she was able to perform simple additions presented by using symbols. When the researchers showed her the cards representing the number 2 and 3, Sheba could answer by pointing to the card identifying the number 5. Considering the fact that the animal could easily perform this task already in the earliest stages of the test, the authors succeeded indeed in proving that for Sheba mathematical additions were not more difficult than the simple action of enumerating objects.

In conclusion, the studies outlined above seem to question the idea that the representative system at the basis of numerical skills is exclusively human. Nevertheless, as surprising as it might seem, it does not mean that chimpanzees, monkeys, apes, mice, dolphins, and birds have the same numerical competences of humans. As we have seen in the case of Sheba, the ability to associate Arabic numerals to the right numerosities is a long and difficult process that requires years of training and, even in this case, results are never completely correct and the right answers come in quite limited number.

As we will see in the next paragraph, the results obtained even in few-month-old children are much more encouraging.

2.3   Numerical Cognition in Children

In the previous paragraph, we outlined some of the evidence corroborating the existence of a representation of "numerosity" in animals. Starting from these considerations, it seems more than plausible to assume that humans are also equipped with a similar system to make representations of quantities. Nevertheless, humans possess arithmetical abilities that go well beyond the rudimental knowledge of animals. With the exception of some individuals affected by specific disorders, the majority of human beings can perceive, express, explain and use the difference between two quantities, whatever these are. On the contrary, as we have outlined, animals struggle to understand lower levels of difference: for example, 8 can be easily distinguished from 16, but we are not sure that the difference between 8 and 10 is even perceived. Therefore, a good question would be: is our arithmetical knowledge related in some way to the system used by animals? Do we have a shared system? And if so, when does this system appear and when do humans learn how to use it?

For a long time, people have supported the idea that children come into this world without any prior knowledge. This hypothesis was considered particularly true for mathematical concepts, since these are quite complex and require an appropriate cognitive development in order to be acquired. Historically, Jean Piaget (1952) was the strongest supporter of this approach. Piaget contended that the acquisition of the concept of number develops through stages that occur in parallel with a process of reinforcement of the structures related to logics. In particular, Piaget believed that the cognitive development of new-borns and children follows a series of stages that leads to the establishment of successive structures. These stages are: 1) the sense-motoric stage (0−2 years), where new-borns interpret the surrounding world by using their senses and actions; 2) the preparatory stage (2−7 years), which represents a pre-functional stage of intuitive reasoning (i.e. reasoning is driven by perceptive intuition); 3) the concrete operations stage (8−11 years), where children can cancel the effect of a concrete action through their reasoning; 4) the formal operations stage (starting from 12 years), where the logical reasoning through abstract propositions, hypotheses, and ideas replaces the reasoning based on concrete objects that characterised the previous stages. Hence, according to the Piaget's constructivist perspective, the acquisition of the concept of number gradually develops over time, from the time of birth to the adult age, one stage after the other, thanks to the interaction between the individual and the surrounding environment.

In particular, in the first stage, children only have a general and intuitive perception of the concept of number; in the second stage they understand the problem and try to combine logics with the still powerful influence of perceptive illusions; in the third stage, finally, they build an awareness of quantities that is kept independently from their physical disposition; in the fourth stage, after discovering the invariability of quantities, children recognise them as something simple and patent, completely detached from any operation or form of reasoning. In a nutshell, children acquire a real form of representation of numbers when they are able to overcome the hurdle represented by the conservation of quantities, something that according to Piaget happens at the age of 7−8 years, when they reach the concrete operations stage.

Piaget's goal was to demonstrate that the establishment of the concept of number does not depend on language (which simply reveals the figurative and static aspects of reasoning), but on the inner action that becomes reversible: in other words, that the concept of number depends on the operational aspects of reasoning that have the scope of performing an action in the world and, through this, understand it. Exercising reasoning requires, according to Piaget, that the conservation of something is assumed as *conditio sine qua non*. For example, it is impossible to dwell on the spatial relations between objects without assuming that these objects keep existing, or reasoning about their weight and length without assuming that their features are permanent. At the same time, Piaget contended that numbers become understandable only when they always keep their essence, independently from the disposition of the units that make them up. This is the reason why the French researcher studied the genesis of numbers in children by resorting to conservation tasks. For example, one of these tasks involved the use of two vases (of identical or different shape) that were gradually filled with some pearls. The content of one vase was then transferred to the other. Piaget noticed that the two-way and mutual relationship between the content of the two vases (which met the wish of the researcher of having a practical form of enumeration) was not sufficient to ensure that children could understand the principle of conservation. As a matter of fact, by relying solely on their perceptive skills, children always indicated the vase where pearls reached the highest level as the vase containing the biggest amount of pearls. Only at the age of seven children started making a correlation between the real equivalence suggested by the comparison procedure and the clear variations of the space occupied by objects. Because of the results of this experiment, Piaget said that it was impossible to find evidence of numerical representation in new-borns.

However, other researchers went against the position taken by Piaget, putting forward the idea that there was a predisposition related to the concept of "quantity" in children since the earliest days of their lives. In 1980, the American psychologists Starkey and Cooper resorted to an experiment centred on the "attention spans" showed by children when faced with something new, in order to understand what they found so surprising and worthy of attention. In their experiment, Starkey and Cooper showed to 4-month-old children pictures containing 2–3 dots for a certain number of times and in succession. Then, they showed them a test image containing 2 or 3 dots. By doing so, Starkey and Cooper proved that, when the numerosity of the image changed (in comparison to the image showed at the beginning of the experiment), the children paid much more attention to the stimulus compared to when the numerosity was the same (Starkey and Cooper 1980).

A couple of years later, Antell and Keating (1983) achieved the same results by following the same experimental protocol but performing the tests on new-borns. In their experiment, four-day-old children were initially shown two cards displaying two black dots, unevenly distributed as regards length and density, until establishing a "habit"; then, the same children were shown a card with three black dots. At this point, it was noticed that new-borns stared at the new card for a time that was more or less 2.5 s longer compared to the other cards. The same happened when the card with three dots was shown before the one with two. This fact did not occur when the new-borns were shown cards containing bigger numerosities (4–6 points). Nevertheless, the authors interpreted these results as a proof of the existence of numerical competences prior to linguistic and counting skills.

At this point, a fair criticism would be that the knowledge observed in children in these experiments does not relate to the concept of number but rather to the concept of numerosity, i.e. the sense related to the size of a group. Nevertheless, this fact does not diminish the importance of this discovery. After all, it is obvious that new-born babies are not able neither to talk nor use linguistic labels to refer to numbers. Nevertheless, even before these experiments, the majority of theories postulated that a sense of the concept of numerosity was developed only after children had learnt to count (and could hence use language). These tests, combined with more systematic ones that we will describe later, prove the contrary, highlighting that even prior to language and counting children are aware of the concept of numerosity.

What could be seriously questioned is the methodological validity of these early studies. The experiments outlined above lack a form of control test on non-numerical parameters. For example, in the experiment by Starkey and Cooper, the authors presented children with non-aligned points, correlating their number to the length and density of the stimulus, but at the same time—without changing the format of numbers—the quantity of matter (i.e. the sum of the area of each dot) increased according to the increase in numerosity. The same goes for the experiment by Antell and Keating, which followed the same experimental protocol. When addressing this criticism, several authors asked whether the positive results were genuinely due to the perception of numerosity or rather to other parameters. From this point of view, we could focus our attention on the experiment by Clearfield and Mix (1999), where researchers—after an initial training stage in which children were shown 2 or 3 dots —tested the reaction of children to new stimuli as far as numerosity was involved, but similar to the previous ones as far as quantity of matter.

In this test, children showed more interest when there was a change in the quantity of matter (with unchanged numerosity), while on the contrary they were completely indifferent to the changes in numerosity when the quantity of matter remained unchanged. In conclusion, Clearfield and Mix emphasized that the data coming from some experiments did not establish beyond all doubts that what drove the reaction of children was indeed numerosity or quantity of matter, suggesting that numerosity is not a salient and easily perceivable feature for children. Nevertheless, three years after his first experiment, Prentice Starkey and colleagues (Starkey et al. 1983)— who were more and more convinced that the representation of numerosity took place in an early stage of development—proved that the number sense in children is not limited to what children see, but that their quantity representation is indipendent from the modality and shared by the visual and auditory system.

By using the paradigm of visual preference, the researchers performed an experiment where two pictures were displayed on a screen: the first one contained two objects, the second one three. Then, an auditory stimulus was introduced (a sequence of two or three drum beats) and it was shown that seven-month-old chil-dren paid more attention to the image with the same number of objects as the number of drum beats they had just heard in the sound sequence. The correlation between auditory sense and visual preference was confirmed also with the quantity of "three". Similar results were achieved with another research methodology by the French psychologist Ranka Bijeljac-Babic and colleagues (Bijeljac-Babic et al. 1993), who based their studies on the suction reflex of new-borns. Babies were given an artificial teat connected to a device that could detect the suction pressure exerted by the baby and played pre-recorded sounds. Starting from the assumption that when the interest of the baby increased, they would have sucked more energetically (and the contrary in the case of a decrease in interest), it was emphasized that the passage from one sound to the other raised the interest of children only when the number of the sounds changed. Several experiments were made to replicate these results in negative terms. For example, David Moore and colleagues (Moore et al. 1987) achieved the opposite result in their experiment: their babies stared longer at the slides containing a dif-ferent number of objects than the number of sounds heard.

In fact, the discrepancy among the data obtained can easily be justified with the substantial methodological differences between the experiments. For example, in Starkey's test the preference shown by children was not immediately clear but appeared only after a certain number of assigned tasks. Moore and colleagues, instead, let the children take a long break between the first and second series of tasks, which could have influenced the answers of babies when they were about to discover the correspondence between the auditory and visual stimuli. Therefore, it is likely that without the break the two experiments would have achieved the same results. Aware of the possibility of alternative explanations, Starkey and some colleagues performed an experiment that trained babies to recognise a specific numerosity through the visual presentation of different images (containing the same number of items). Then, the researchers played for the babies a sequence of 2 or 3 noises and, at the same time, hid the images behind a screen. The outcomes of this experiment proved that babies paid more attention to a situation in which the number of noises was equal to the number of items that they had seen during training (Starkey et al. 1990). Furthermore, the ability to draw intermodal comparisons of small numerosities already at the early age of 7 months has been confirmed by another study by Tessey Kobayashi and colleagues (Kobayashi et al. 2005).

Moreover, starting from 2000, some studies have been carried out in order to investigate big numerosities. An example is the experiment performed by Xu and Spelke (2000) in which six-month-old babies were shown, in a preliminary stage of the test, a series of images containing 8 or 16 dots with changing format and disposition. Subsequently, the same babies were shown new images containing new numerosities (16 dots if in the preliminary stage they had seen 8, and 8 if they had seen 16) and their format was calculated to make sure that the quantity of matter did not change in relation to the images shown in the preliminary stage. In this experiment, the babies stared for a longer time and with more attention only at a new numerosity. Xu and Spelke reproduced this experimental protocol in another test with numerosities nearer to each other, such as 8 dots *vs.* 12 dots, noticing that six-month-old babies did not pay so much attention in this case.

As we have previously noticed when addressing numerical abilities in animals, the quotient (ratio) between two numbers represents the crucial factor when predicting the possibility of distinguishing two numerosities. When faced with similar experimental settings, six-month-old babies were able to distinguish 16 dots from 32, but mixed up 16 with 24. Children are hence able to distinguish bigger numerosities on the condition that their quotient is big enough. By focussing on older children (nine-month-old), Xu and Arriaga (2007) showed that, contrary to what happened in six-month-old babies, they could distinguish numerosities in a 2:3 ratio. These outcomes were confirmed by a research on auditory modalities performed by Lipton and Spelke (2003). In their experiment, the researchers showed that six-month-old babies are able to distinguish the numerosities 8 and 16 (ratio 1:2) but that they are not able to distinguish the numerosities 8 and 12 (ratio 2:3). On the contrary, at the age of nine months, children could distinguish 8 and 12, but not 8 and 10 (ratio 4:5). The extreme similarities of these performances with those observed in visual experiments can lead to the conclusion that children use the same abstract representation of numerosity in both cases.

These results reveal the existence in new-borns of an approximate numerical system that gets more accurate over the first few months of life, developing at an earlier stage than the learning of exact symbolic competences, and they also highlight an important change occurring between the age of 6 and 9 months (the period when linguistic understanding starts taking shape). This fact can be deducted from a study performed by Wood and Spelke (2005) in which six-month-old babies were not able to recognise the difference between a sequence of 4 jumps vs. 6 jumps made by a doll, while the difference was recognisable by nine-month-old children.

2.3.1   Karen Wynn's Dolls

The most striking demonstration that children can indeed perform simple operations of addition and subtraction was made by the American researcher Karen Wynn (Wynn 1990; 1992) through a series of experiments that exploited, also in this case, the ability of children to recognise physically impossible events by staring at them more intensively: children normally follow the objects presented to them and adapt their attention whenever new objects are added or subtracted. In her experiment, five-month-old babies looked at a doll theatre equipped with a screen that could be rolled up and down. The theatre was initially empty, then the researcher placed a doll in it. The screen was then rolled down, hiding the scene, and a second doll was introduced. At this point, the screen was removed and the children were presented with two dolls. The sequence was repeated several times, but in some instances the dolls shown to the children represented impossible results (for example, $1 + 1 = 3$ or $1 + 1 = 1$). In these cases, the children stared at the scene for a longer time compared to when two dolls were shown. Wynn achieved the same results also after that the experimental procedure was changed in order to test the children's skills in understanding subtractions (Fig. 2.3).

In the second experiment, two dolls were initially shown on the screen and then one of them was taken out $(2-1)$. Also in this case, the impossible result $(2 - 1 = 2)$ entailed a longer staring time compared to the right result $(2 - 1 = 1)$.

Therefore, despite the fact that in this experiment the right number of objects was the main driver of the babies' behaviour rather than an approximate distinction (e.g. one doll vs. several dolls), in both tested conditions Wynn came to the conclusion that, at the end of the experiment, there was a higher level of interest by babies only when the final result presented a numerical oddity. It is important emphasizing that this experiment, which aimed at showing that children could make simple arithmetical calculations, led to the development of at least three possible theories that tried to explain the observed behaviours. The first one, outlined by Wynn herself, supported the idea of the existence of an abstract representation of numerosity that allows children to have a representation of each quantity and perform on these quantities mental operations.

A second explanation was put forward by Alan Leslie and colleagues (Leslie et al. 1998) and by Tony Simon (1999), taking inspiration from the theory of "object files". According to this theory, an object is identified by using "files" that allow us to follow objects in their movement and link them to different perceptions, spread over time and space. Children, therefore, have a form of inborn physical understanding that has the duty of providing the idea of the permanence of objects (even when they are momentarily hidden). Babies follow indeed the objects presented to them and adapt their attention accordingly when new objects are added or taken. Furthermore, according to the object files theory, babies can infer some kind of numerical information starting from the physical features of the objects' trajectories: in this way, when two objects are presented to them one after the other, they do not pay the same attention in the case that the two objects are hidden behind

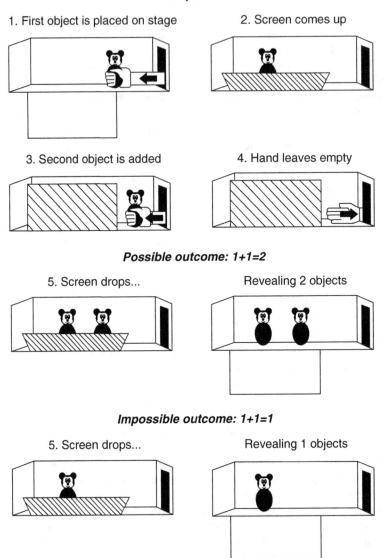

Fig. 2.3 Representation of the procedural stages of the experiment where right and wrong arithmetical additions are shown to five-month-old babies (Wynn 1992)

the same screen or behind two different screens. In the latter case, babies know that behind the screens there are two objects, while in the former case they expect that there is only one object (Spelke et al. 1995). The number of files simultaneously available is limited to 4 and this explains why it is impossible to follow simultaneously more than 4 moving objects belonging to the same group. This is the reason why after adding 3 or 4 objects, it is impossible to add another item, because otherwise children would lose access to numerosity, due to lack of other files (Feigenson et al. 2002).

Starting from files, it is hence possible to infer the numerosity of groups of objects (e.g. two groups of objects can be compared using a term-to-term correspondence between files) and make simple operations, but they do not allow the representation of numerosity. According to Simon (1999), the babies involved in Wynn's experiment did indeed use the files' system to follow the objects on the scene, and activating an additional file to follow the new object. In this way, babies seemed sensitive to numerosity without having a clear representation of it. It is useful though to emphasize that this limit is not incurred in experiments performed with separable objects.

An example is the study carried out by Wynn and colleagues (Wynn et al. 2002), where videos depicting different groups of dots were displayed on a screen to a group of children. In the first stage of the experiment, two (or 4) groups of 3 dots moved independently on the screen. Then, children were tested with two stimuli: 2 groups of 4 dots or 4 groups of 2 dots. In both cases, the stimuli had the same number of dots and, therefore, they were identical with the exception of the number of groups presented. In the end, children were able to distinguish the two different kinds of stimulus, showing that they did not encounter difficulties in dealing with numerosities when the calculation passages were not directly associated to specific objects.

Finally, the third theory was developed by Cohen and Marks (2002), who interpreted the data obtained by Wynn as the result of a "low profile" process, i.e. the proof of the preference of babies for the latest stimuli presented to them. According to the two researchers, babies stared for a longer time to the result "1" in the setting "1 + 1=2 *vs.* 1" not because they were surprised, but rather because they had seen one doll at the beginning of the experiment and therefore they were attracted to it when it reappeared on the scene. Similarly, babies preferred watching two dolls in the setting "2 − 1 = 1 *vs.* 2" because at the beginning of the experiment they had seen two dolls. To check their hypotheses, Cohen and Marks made an experiment that followed the experimental protocol developed by Wynn but presented possible results of 0, 1, 2, or 3 dolls. In this experiment the authors reached the same results of Wynn in the case of 1 and 2 dolls (the babies stared for a longer time at 1 in the case of "1 + 1" and more at 2 in the case of "2 − 1"), nevertheless the same result was not obtained for 0 and 3. It is though necessary to highlight some procedural features that question the soundness of these results. The experiment by Cohen and Marks, in fact, multiplies the experimental conditions and the preliminary tests, so much so that it is plausible to believe that babies were tired even before starting the test. Furthermore, the tests providing numerical coherent results were very few (almost 25%).

Despite the criticism outlined above, Wynn's hypothesis was supported by several experiments that showed the existence in children of a representation system of numerosities not only in the case of small quantities, but also in the case of bigger numerosities. McCrink and Wynn (2004) performed an experiment where five-month-old babies underwent a test of approximate addition and subtraction with numerosities 5 + 5 and 10 − 5. In this test, some objects were presented to the babies by keeping them once again behind a screen and by changing format, in order to avoid children getting interested in physical indexes (for example, the format of dots, the quantity of matter, etc.). Once the objects were hidden behind the screen, the researchers added new objects (addition), or hid some of these objects behind the screen (subtraction). In both cases, when the screen was lowered and the two results were presented (5 and 10), it was noticed that children paid more attention when the result was incorrect.

Stanislas Dehaene and his colleagues (Piazza et al. 2004) are the authors who delivered the most striking demonstration of the continuity of representation from smaller to bigger numerosities. By using the recording of the correlated event potentials, the researchers tested babies of an age comprised between 94 and 124 days both on big numerosities (4, 8, and 12) and small numerosities (2 and 3). In particular, in the first stage of the experiment, the babies were shown an initial image containing from 2 to 5 stimuli that lasted for 1500 ms and, after establishing this habit, they were presented with a second image containing a new numerosity. To be sure that the children's answers were based on the numerosity of stimuli and not on other parameters, all non-numerical parameters were kept under strict control, such as the lightness of stimuli or their total occupied area. In this way, three pairs of numerosities got tested: big numerosities with a big distance (4 vs. 12), big numerosities with a small distance (4 vs. 8), and small numerosities (2 vs. 3). In the second stage of the experiment, the numerosities were presented using auditory stimuli in place of the visual ones used previously.

The data obtained by analysing the brain activities in babies showed the same effect for all series of tests, performed using auditory and visual stimuli, and presented both in sequence or simultaneously on small and big numerosities. The results emphasized that babies made numerical representations of a high level of abstraction for all different formats of stimuli and without any difference between small and big numerosities. Therefore, contrary to what was proven by other behavioural data (Feigenson et al. 2004), the results obtained by Dehaene suggest that children have access to numerical representations both for smaller and bigger numerosities.

Yet, how can the divergence from behavioural results be explained? According to Lisa Feigenson, an explanation can be found in the fact that, for stimuli relating to smaller numerosities, babies focus more on other features of the stimuli, such as the quantity of matter. Despite the fact that babies are able to make abstractions related to numerosity, these do not have a behavioural effect. According to this interpretation, Feigenson showed that in some experimental settings, numerosity may become a salient feature for babies even in small quantities such as 2 or 3 (Feigenson 2005). Obviously, this does not mean that there is not a development in the accuracy of numerical representations in children: it has indeed been proven that six-month-old babies are able to distinguish two distant numerosities with a ratio of 2 (such as 8 and 16, 16 and 32), but fail with a ratio of 1.5 (8 and 12, 16 and 24). Nine-month-old babies instead have the ability to distinguish 8 and 12, but fail at doing so with nearer numerosities (8 and 10).

Elizabeth Brannon (2002) provided another evidence of the fact that children have competences related to the recognition of numerosities and counting already after the first year of life. In an experiment that used the habit paradigm, Brannon showed that starting from 11 months, babies understand the concept of ordinal numbers, since they can understand the difference between numerosities put in increasing or decreasing order. In her experiment, 9-month-old and 11-month-old babies were shown images depicting groups of dots in increasing or decreasing order. Then, a test sequence presenting a new (either increasing or decreasing) numerosity was introduced. In 9-month-old babies, Brannon did not notice any kind of particular reaction to the two sequences, while in the group of 11-month-old babies she recorded an increased interest in the case of the sequence presented in a new order.

2.4 The Approximate Number System

In spite of the criticism presented to the studies outlined above and the fact that the theoretical debate is still on-going, the research in this field paved the way to a new generation of studies on numerical cognition that provided deeply interesting insights. In particular, authors such as Stanislas Dehaene and Brian Butterworth launched new models of numerical cognition that tried to provide an answer to the ancient debate nature *vs.* culture, inborn *vs.* learnt, aiming at conciliating Piaget's ideas with the most recent experimental data. Their combination was made possible by understanding numerical competence as something affected both by culture and linguistic learning, as Piaget contended, and characterised by some inborn skills, as recent experiments emphasized. Dehaene, for example, developed a model were two different systems of numerical representation are postulated: an approximate analogical system that allows for an approximate and inaccurate counting of numerical quantities, and an accurate and symbolic system.

The former system is culture- and language-independent and is made possible by an organ in the brain that is designed to perceive and make a representation of numerical quantities, features that link it to the pro-arithmetical skills shown by new-borns and animals. On the contrary, the latter system depends on culture and the learning of symbols and rules, hence it is strictly connected to language, becoming a typical feature of adults. Dehaene postulated that humans are equipped with a "mathematical sense" that they share with other animal species, an instinct that represents the expression of the functioning of a "mental organ", a set of cerebral wires that belongs also to other species and works as a form of approximate counting device. This "device" allows us to perceive, memorise, and compare numerical quantities (Dehaene 1992). According to the accumulator system introduced by Meck and Church (1984) to account for the arithmetical skills of mice, each object is represented by the central nervous system as an impulse. Subsequently, the cognitive system accumulates the impulses produced after seeing each stimulus, transferring the information to the long-term memory, which is tasked with the general labelling of impulses.

To better understand how the accumulator works, it is useful to mention the water reservoir metaphor described by Dehaene (1997). According to the French neuroscientist, we must imagine that each countable unit is represented by a quantity of water that is added to a reservoir. By taking stock of the water level in the reservoir, it is possible to compare the different sizes of reservoirs. At the same time, additions or subtractions are possible by simply adding or taking out water. The accumulator works by recording events; a "drop of water" for each event. In this way, different levels represent different numbers. Nevertheless, since this system cannot represent accurately the exact level of impulses, its functioning is affected by the distance effect and the size effect. In fact, the lower the difference between two sets is, the harder it is to compare them (distance effect). At the same time, the bigger the size of the sets, the harder it is to distinguish them (size effect). The accumulator does not succeed in providing accurate results because it cannot accurately represent the level of impulses. Nevertheless, compared to other mechanisms such as object files, the accumulator allows for the comparison of very big numerosities, provided that there is enough numerical distance between the two analysed groups.

The accumulator it refers to the technical concept of numerosity, i.e. the simple perceptual estimation of different sets of objects and the ability to understand whether two sets of objects are equal or not. This ability has been found both in infants and animals and it is called 'protonumerical skill' or 'pre-numerical skill'.

Starting from these considerations, over the last two decades, neuroscientists like Dehaene and many others developmental psychologists carried out a series of studies with the aim of finding the 'anti-Piagetian' empirical proof that human beings are born with a 'number sense'.

As Dehaene explains in the preface to the second edition of his book, *The Number sense:*

> Fifteen years have elapsed since I proposed my number sense hypothesis — the peculiar idea that we owe our mathematical intuitions to an inherited capacity that we share with other animals, namely, the rapid perception of approximate numbers of objects. How does such a preposterous notion hold up after fifteen years of intense scrutiny? Surprisingly well, I would say. Number sense is now recognized as one of the major domains of human and animal competence, and its brain mechanisms are constantly being dissected in increasing detail. (Dehaene 2011, p. 237).

Conversely accumulator, the second system is based on symbols and it is language- and culture-dependent; it is typical of adults; and it is founded on the ability of counting, therefore on a numerical system and on all arithmetical operations.

The awareness of the cultural nature of exact arithmetic is due, according to Dehaene (2011) to the 'courage' and talent of some anthropologists and linguists 'who took the pains to travel great distances in order to investigate the mathematical competence of remote cultures' in which there is a minimal mathematical vocabulary that in most cases only includes the words for 'one', 'two', 'three', 'a lot'. In the second chapter, I will present the main ideas on this topic.

References

Antell, S. E., & Keating, L. E. (1983). Perception of numerical invariance by neonates. *Child Development, 54,* 695–701.

Beran, M. J., & Beran, M. M. (2004). Chimpanzees remember the results of onebyone addition of food items to sets over extended time periods. *Psychological Science, 15,* 94–99.

Bijeljac-Babic, R., Bertoncini, J., & Mehler, J. (1993). How do 4-day-old infants categorize multisyllabic utterances? *Developmental Psychology, 29*(4), 711–721.

Biro, D., & Matsuzawa, T. (2001). Use of numerical symbols by the chimpanzee (Pan troglodytes): Cardinals, ordinals, and the introduction of zero. *Animal Cognition, 4,* 193–199.

Boysen, S. T., & Berntson, G. G. (1989). Numerical competence in a chimpanzee. *Journal of Comparative Psychology, 103*(1), 23–31.

Brannon, E. M. (2002). The development of ordinal numerical knowledge in infancy. *Cognition, 83,* 223–240.

Brannon, E. M., & Terrace, H. S. (1998). Ordering of the numerosities 1 to 9 by monkeys. *Science, 282,* 746–749.

Brannon, E. M., & Terrace, H. S. (2000). Representation of the numerosities 1-9 by rhesus macaques (macaca mulatta). *Journal of Experimental Psychology: AnimalBehavior Processes, 26*(1), 31–49.

Cantlon, J. F., & Brannon, E. M. (2006). The effect of heterogeneity on numerical ordering in rhesus monkeys. *Infancy, 9,* 173–189.

Clearfield, M. W., & Mix, K. S. (1999). Number versus contour length in infants' discrimination of small visual sets. *Psychological Science, 10*(5), 408–411.

Cohen, L. B., & Marks, K. (2002). How infants process addition and subtraction events. *Developmental Science, 5*(2), 186–201.

Davis, H., & Albert, M. (1987). Failure to transfer or train a numerical discrimination using sequential visual stimuli in rats. *Bulletin of the Psychonomic Society, 25,* 472–474.

Davis, H., & Perusse, R. (1988). Numerical competence in animals: Definitional issues, current evidence and a new research agenda. *Behavioural Brain Science, 11,* 561–579.

Dehaene, S. (1997). *The number sense.* New York, Cambridge (UK): Oxford University Press, Penguin Press.

Dehaene, S. (1992). Varieties of numerical abilities. *Cognition, 44,* 1–42.

Dehaene, S. (2011). *The number sense. How the mind creates mathematics. Revised and updated edition.* New York: Oxford University Press.

Devlin, K. (2005). *The math instinct. Why you're a mathematical genius.* New York: Thunder's Mouth press.

Emmerton, J., & Delius, J.D. (1993). Beyond sensation: Visual cognition in pigeons. In H. P. Zeigler & H. Bischof (Eds.), *Vision, brain, and behavior in birds* (pp. 377–390). Cambridge, MA: MIT Press.

Emmerton, J., Lohmann, A., & Niemann, J. (1997). Pigeon's serial ordering of numerosity with visual arrays. *Animal Learning and Behavior, 25,* 234–244.

Feigenson, L. (2005). A double-dissociation in infants' representations of object arrays. *Cognition, 95,* 37–48.

Feigenson, L., Carey, S., & Hauser, M. D. (2002). The representations underlying infants' choice of more: Object files versus analog magnitudes. *Psychological Science, 13*(2), 150–156.

Feigenson, L., Dehaene, S., & Spelke, E. S. (2004). Core systems of number. *Trends in Cognitive Sciences, 8,* 307–314.

Fernandes, D. M., & Church, R. H. (1982). Discrimination of the number of sequential events by rats. *Animal Learning and Behavior, 10*(2), 171–176.

Flombaum, J. I., Junge, J. A., & Hauser, M. D. (2005). Rhesus monkeys (Macaca mulatta) spontaneously compute addition operations over large numbers. *Cognition, 97,* 315–325.

Gallistel, C. R., & Gelman, R. (2000). Non-verbal numerical cognition: From reals to integers. *Trends in Cognitive Sciences, 4,* 59–65.

Hanus, D., & Call, J. (2007). Discrete quantity judgments in the great apes (Pan paniscus, Pan troglodytes, Gorilla gorilla, Pongo pygmaeus). *Journal of Comparative Psychology, 121*(3), 241–249.

Hauser, M. D., & Spelke, E. S. (2004). Evolutionary and developmental foundations of human knowledge: A case study of mathematics. In M. Gazzaniga (Ed.), *The cognitive neurosciences* (Vol. 3). Cambridge: MIT Press.

Jaakkola, K., Fellner, W., Erb, L., Rodriguez, A. M., & Guarino, E. (2005). Understanding of the concept of numerically "less" by bottlenose dolphins (Tursiops truncatus). *Journal of Comparative Psychology, 119*, 296–303.

Jordan, K. E., & Brannon, E. M. (2006). Weber's law influences numerical representations in rhesus macaques (Macaca mulatta). *Animal Cognition, 9*, 159–172.

Kilian, A., Yaman, S., Fersen, L., & Gunturkun, O. (2003). A bottlenose dolphin (Tursiops truncates) discriminates visual stimuli differing in numerosity. *Learning and Behaviour, 31*, 133–142.

Kilian, A., Von Fersen, L., & Güntürkün, O. (2005). Left hemispheric advantage for numerical abilities in the bottlenose dolphin. *Behavioural Processes, 68*, 179–184.

Kobayashi, T., Hiraki, K., & Hasegawa, T. (2005). Auditory-visual intermodal matching of small numerosities in 6-month-old infants. *Developmental Science, 8*, 409–419.

Leslie, A., Xu, F., Tremoulet, P., & Scholl, B. (1998). Indexing and the object concept: Developing what and where systems. *Trends in Cognitive Science, 2*(1), 10–18.

Lipton, J. S., & Spelke, E. S. (2003). Origins of number sense: Large-number discrimination in human infants. *Psychological Science, 14*(5), 396–401.

Lyon, B. E. (2003). Egg recognition and counting reduce costs of avian conspecific brood parasitism. *Nature, 422*, 495–499.

Marler, P., & Tamura, M. (1962). Song dialects in three populations of Whitecrowned sparrows. *Condor, 64*, 368–377.

McComb, K., Packer, C., & Pusey, A. (1994). Roaring and numerical assessment in the contests between groups of female lions, Panther leo. *Animal Behaviour, 47*, 379–387.

McCrink, K., & Wynn, K. (2004). Large number addition and subtraction by 9-month-old infants. *Psychological Science, 15*(11), 776–781.

Mechner, F. (1958). Probability relations within response sequences under ratio reinforcement. *Journal of the Experimental Analysis of Behaviour, 1*, 109–122.

Mechner, F., & Guevrekian, L. (1962). Effects of deprivation upon counting and timing in rats. *Journal of the Experimental Analysis of Behavior, 5*(4), 463–466.

Meck, W. H., & Church, R. M. (1984). Simultaneous temporal processing. *Journal of Experimental Psychology: Animal Behavior Processes, 10*(1), 1–29.

Moore, D., Benenson, J., Reznick, J. S., Peterson, M., & Kagan, J. (1987). Effect of auditory numerical information on infants' looking behavior: Contradictory evidence. *Developmental Psychology, 23*, 665–670.

Pepperberg, I. (2006). Grey parrot numerical competence: A review. *Animal Cognition, 9*, 377–391.

Piaget, J. (1952). *The child's conception of number.* New York: Norton.

Piazza, M., Izard, V., Pinel, P., Bihan, D. L., & Dehaene, S. (2004). Tuning curves for approximate numerosity in the human intraparietal sulcus. *Neuron, 44*(3), 547–555.

Platt, J. R., & Johnson, D. M. (1971). Localization of position within a homogeneous behaviour chain: Effects of error contingencies. *Learning and Motivation, 2*, 386–414.

Robins, A., Lippolis, G., Bisazza, A., Vallortigara, G., & Rogers, L. J. (1998). Lateralized agonistic responses and hind-limb use in toads. *Animal Behaviour, 56*, 875–881.

Rugani, R., Regolin, L., & Vallortigara, G. (2007). Rudimental numerical competence in 5-day-old domestic chicks: Identification of ordinal position. *Journal of Experimental Psychology: Animal Behavior Processes, 33*(1), 21–31.

Rumbaugh, D. M., Savage-Rumbaugh, S., & Hegel, M. T. (1987). Summation in the chimpanzee (pan troglodytes). *Journal of Experimental Psychology: Animal Behavior Processes, 13*, 107–115.

Santos, L. R., Barnes, J. L., & Mahajan, N. (2005). Expectations about numerical events in four lemur species. *Animal Cognition, 8*, 253–262.

Simon, T. J. (1999). The foundations of numerical thinking in a brain without numbers. *Trends in Cognitive Sciences, 3*(10), 363–364.

Spelke, E. S., Kestenbaum, R., Simons, D. J., & Wein, D. (1995). Spatio-temporal continuity, smoothness of motion and object identity in infancy. *British Journal of Developmental Psychology, 13,* 113–142.

Starkey, P., & Cooper, R. G. (1980). Perception of numbers by human infants. *Science, 210,* 1033–1035.

Starkey, P., Spelke, E. S., & Gelman, R. (1983). Detection of intermodal numerical correspondences by human infants. *Science, 222*(4620), 179–181.

Starkey, P., Spelke, E. S., & Gelman, R. (1990). Numerical abstraction by human infants. *Cognition, 36,* 97–128.

Sulkowski, G. M., & Hauser, M. D. (2001). Can rhesus monkeys spontaneously substract? *Cognition, 79,* 239–262.

Uller, C., Hauser, M. D., & Carey, S. (2001). Spontaneous representation of number in cotton-top tamarins. *Journal of Comparative Psychology, 115,* 1–10.

Uller, C., Jaeger, R., Guidry, G., & Martin, C. (2003). Salamanders (Plethodon cinereus) go for more: Rudiments of number in a species of basal vertebrate. *Animal Cognition, 6,* 105–112.

Wilson, M. L., Hauser, M. D., & Wrangham, R. W. (2001). Does participation in intergroup conflict depend on numerical assessment, range, location, or rank for wild chimpanzees? *Animal Behaviour, 61,* 1203–1216.

Wood, J. N., & Spelke, E. S. (2005). Infants' enumeration of actions: numerical discrimination and its signature limits. *Developmental Science, 8,* 173–181.

Woodruff, G., & Premack, D. (1981). Primative (sic) mathematical concepts in the chimpanzee: Proportionality and numerosity. *Nature, 293,* 568–570.

Wynn, K. (1990). Children's understanding of counting. *Cognition, 36,* 155–193.

Wynn, K. (1992). Addition and subtraction by human infants. *Nature, 358,* 749–750.

Wynn, K., Bloom, P., & Chiang, W. (2002). Enumeration of collective entities by 5-month-old infants. *Cognition, 83,* B55–B62.

Xu, F., & Arriaga, R. I. (2007). Number discrimination in 10-month-old infants. *British Journal of Developmental Psychology, 25,* 103–108.

Xu, F., & Spelke, E. S. (2000). Large number discrimination in 6-month-old infants. *Cognition, 74,* 1–11.

Chapter 3
The System 2

Abstract We think that the numbers are accurate, that tells us the truth, and that are not exposed to any interpretation. Consequently, the numbers are merely numbers, regardless of the context in which they occur, regardless of the culture of single population. However, in recent years there have been some studies that go against this conception of common sense. This chapter considers the role of language, focusing in particular on studies of Munduruku and Chinese numerical cognition.

Keywords Language and numbers · Number representation · Culture
Ethnomathematics

3.1 Numbers and Colours

Everyone believes that numbers are accurate, truthful, and cannot be interpreted. Therefore numbers are for many just numbers, no matter the context or the culture of the population using them. Nevertheless, there is evidence that goes against this common sense approach. Even in their plainest use, numbers acquire different meanings, particularly as far as the words used to identify them are concerned. For example, there are clear differences between the use of numbers in schools and their use in daily life problems, such as mathematical issues encountered at the supermarket. These differences are due to the fact that problems apparently similar in nature can be perceived or tackled differently when they appear in different contexts (Lave 1988; Nunes et al. 1993). It could seem logical that there is a substantial difference between the use of numbers in "pure calculations" (such as performing the mental calculation $180:3 = 60$) and the use of numbers in modalities with clear significants (as for poetic calligraphy or in slogans using numbers). If numbers may look like elements that are relatively language-independent in the first scenario, in the second one it is clear that they strongly depend upon it. In other words, numbers acquire another meaning when there are good reasons to mull over them, when they have an important meaning (such as the amount of calories or profits).

© The Author(s) 2018   39
M. Graziano, *Dual-Process Theories of Numerical Cognition*,
SpringerBriefs in Philosophy, https://doi.org/10.1007/978-3-319-96797-4_3

These considerations lead us to the conclusion that numbers are, at the end of the day, quite similar to words. They are full of meaning and we think about them similarly to how we think about other items. After this introduction, in this chapter we will deal with the significant influences that some languages have on the learning of mathematics. Our goal is not to justify different mathematical skills by resorting to cultural arguments and cultural differences, but rather to highlight the differences that can exist between cultural groups in the learning of mathematics and show that there are relevant cultural aspects worthy of attention in this discussion. In fact, for many years, the dominating idea (particularly in linguistics) has been that categories do not exist in the external world, nor have a perception basis: reality has been considered as an indefinite *continuum* and the fact that humans tend to categorise elements is ultimately nothing more than a convention stemming from the learning process. Initially, philosophers assumed that names could be given only to things that humans could see and, according to a questionable *modus tollens*, if a dictionary did not contain a certain term, the corresponding object was not perceivable. The names used to designate colours became immediately the privileged battlefield for competing linguistic theories: the relationship existing between "perceptions-language" has been debated for more than 80 years. Edward Sapir put forward the idea of linguistic relativity, which was then developed in the 1950s by Benjamin Whorf. The main argument supported by this theory is that language affects or at least shapes reasoning and cognition. In particular, Whorf contended that there is no systematic or universal way to categorise data, but rather linguistic factors that affect the way in which humans make cut-offs of nature, organise concepts, and ascribe a meaning to them (Whorf 1956). This happens at different levels, as the following example shows. In Greenland, the Inuit people speaks a language that allows them to identify more than twenty different types of snow, while in Italy we just have one word for snow. This could lead us to the conclusion that, by having more than twenty different ways to identify snow, the perception of children that have grown up in such an environment is quite different to that of Italian children. In a snow-covered environment, Inuit children would identify different categories of snow where others would just see a pristine white surface. According to a second idea, the influence of language is perceivable at a higher level, i.e. at the level of conceptual categories. As Inuits do, Italians could also perceive the difference between two types of snow, a difference related to two different words. The problem is that Italians categorise all their perceptions related to snow under the same concept and therefore the two types of snow seem equivalent to them. Whatever type of snow they see, they interpret it in the same way, inferring the same information. On the contrary, Inuits infer from the same event a series of different expectations (how the snow will react when walking on it, how much it will last before melting, at which temperature it will melt, etc.). It is also possible to imagine a scenario where concepts are acquired independently of language. This is the idea supported by Stevan Pinker. In his work "The language instinct", Pinkler disputes ideas such as "Inuits in Greenland have thirty words to name different kinds of snow while we only have one" or "Inuits can recognise thirty different snow qualities, we cannot!" statements that often appear in academic papers and journals, highlighting in a methodical and systematic way that having at disposal a wide range of words is a common feature of all languages spoken in continental climate regions (for example, Maoris in New Zealand can distinguish different nuances of red). Pinker contends that language is merely a tool to convey one's ideas and understand others'. The richness of vocabulary of continental languages to identify snow simply reflects the stronger need to communicate ideas related to this issue for these populations compared to other cultures.

As it is well known, the relativist stance has been criticised by Berlin and Kay (1969), who—commenting about the terms used to identify colours—contended the existence of a universal vocabulary that does not depend on the language used. To confirm their theory, the authors studied almost twenty languages, noticing that beyond a superficial difference, there are recurring patterns among languages: possible colour categories are limited (from 2 to 11 basic colours) and the establishment of colour names—in line with this non-linguistic category—follows a set of strict limitations. Following this research, other studies on the categorisation of colours were carried out by comparing very different cultures and languages: for example, the study on categorisation involving Americans and the Dani people of Java (Rosch-Heider 1972), and the studies focusing on American English and Japanese (Uchikawa and Boynton 1987) or on English, Russian, and Setswana (Davies and Corbett 1997).

The WCS (*World Color Survey*), a study performed on almost 110 languages, came to three conclusions: (1) there are some substantial trans linguistic groups (categories) that gravitate around some privileged points of the colour spectrum, (2) these privileged points are similar both in the oral languages of non-developed communities and the written languages of developed societies, and (3) these privileged points are placed in proximity to the colours named red, yellow, green, blue, purple, brown, orange, pink, black, white, and grey (Kay and Regier 2003). How is it possible to explain that colours are perceived as the same even if different languages are used? According to linguists such as Chomsky or Pinker, all humans share the same *mentality*, i.e. a "language of the mind" that connects the processing of information at neuronal level and spoken language, therefore the deep structure of grammar is not only universal but also inborn. Hence, at a conceptual level, the effect of language will always be to drive the attention towards the environment rather than towards other factors.

By talking about colours instead of numbers, we do not want to create more confusion, but rather emphasise that the case of colours is often indicated as the best and most symbolic example of a persistent feature of physical objects. Frege, for example, in *Foundations* wrote that numbers and colours have similarities (for example, both are objective), but in the end only colour represents a perceivable property of objects. The German philosopher contended that a specific colour belongs to a surface regardless of our judgment, while a number belongs to an object according to how we consider the object at hand and, therefore, only in relative terms. In addition to this, numbers also differ from colours under a much more obvious aspect than Frege's arguments: when we talk about colours we talk about blue, red, and yellow, coloured objects, or combinations of colours (e.g. orange or pink). On the contrary, talking about numbers does not mean talking about 5, 20, and 45, numbered items or some combinations of numbers. Furthermore, the number of discernable colours is limited (humans are able to distinguish up to ten million colours associated with the light streams produced by light sources and 7.5 million colours in the case of reflected light), while numbers are virtually infinite. Frege's considerations underlined the fact that if numbers were mere mental representations, we should drastically limit their domain, because our ability of representation is limited and the existence of infinite numbers is thus impossible. Furthermore, Frege fretted that if thoughts were private and belonged only to their authors as much as representations belong only to those who make them, then two people with different thoughts on the simplest arithmetical truth would not fall in contradiction. In other words, his doubts could be summed up by the question: different minds, same numbers?

At the same time, competent counterparts can easily convey the truths of the theory of numbers that they detain thanks to the explicitly infinite nature of the mathematical language. But what would happen if these counterparts did not have such a rich and articulated language as ours? In this case, could a numerical system worthy of the name even exist? These are the questions that Peter Gordon tried to answer. This professor of bio-behavioural sciences at the *Columbia University* of New York decided to travel to the Brazilian Amazon Rainforest in order to study the Pirahã population, whose language lacks almost completely names for numbers and whose system is axed on the categories "one-two-many" (Gordon 2004). The author wished to experimentally verify the extraordinary observations made by another linguist and anthropologist, Dan Everett, professor of phonetics at the University of Manchester, who lived for decades among the Pirahã and hosted Gordon himself.

In a study that stirred strong reactions in the scientific world, Gordon started his analysis by verifying that the only available names for numbers to the Pirahã population were the equivalents to "one", "two", and "many", with the term "one" that should be considered as "almost one", the term "two" as "a little bit more than one" or "few", and all remaining numbers that fell into the category identified by the word "many". Once this fact was established, he discovered that the words for 1 and 2 were polysemic and could be loosely used to quantify different situations. Furthermore, there was a striking phonetic similarity between the words used to identify "one": *hói* and "two": *hoí*. Gordon proceeded to test some individuals belonging to the tribe, proving their inability to accurately use fingers to indicate small quantities. For example, they could not group together familiar objects (small sticks or nuts) to replicate the sets created by the researcher, or use counting strategies based on unit repetition. When faced with tasks requiring the replication of combinations of continuous series of a small number of familiar objects or with simple tasks involving the reconstruction of series that required visual memory or spatial orientation skills, the individuals tested by Gordon achieved performances in line with the limits set by the almost complete lack of numerical language in their culture: they were successful for the first three numbers, while their performances deteriorated with numbers over 3. Therefore, the skills linked to counting up to three *items* were stable. Hence, Gordon contended that his results soundly confirmed the Whorfian hypothesis that thoughts are affected by language and the words used to convey ideas. As the author explained, one can have a sense of "three items" without having a word to express it, but cannot have a sense of "four or more items" without having words to express it.

After publishing the results of his research in the journal *Science*, Gordon was criticised from two points of view. Surprisingly enough, the first criticism came from Everett. The linguist rejected the idea that these results obtained on the field confirmed a simple stance of linguistic determinism. He believed that it was the Pirahã's culture in general that presented a series of substantial peculiarities. Everett contended that the Pirahã's linguistic culture is limited to non-abstract arguments that relate to an immediate experience. This limitation explained the high number of surprising elements in Pirahã's grammar and culture, such as the absence of legends about the creation of the world, the simplest familiar system in the world, the absence of enumeration forms and concepts of calculation, the absence of terms to denote colours, the absence of all quantification terms, etc. (Everett 2005). In other words, Dan Everett believed that the Pirahã "could not count because they did not want to".

The second critical observation was presented by several researchers and was addressed both to Everett and Gordon. In particular, the two authors were criticised for willingly ignoring the methodological question in assessing contexts of use for quantification and enumeration operations and their possible social inutility. Going into details, critics meant that it was difficult to obtain reliable results on the general "Pirahã enumeration" when the preliminary tests on the possible existence of a numerical calculation system had involved only two individuals. Furthermore, the main research tests were submitted to seven individuals (six men and one woman) from two different villages and "most of the data came from four of these men that were particularly willing to participate in the research" (Gordon 2004, p. 497). Therefore, Gordon's results, which were "driven" by Everett, in addition to reservations on their representativeness, must only be seen as a selected sample and cannot be deemed as indicative.

3.2 Ethnomathematics Studies

As for colour names, also mathematics should have a basic mathematical vocabulary that includes the words for "one", "two", and "many". Some languages do not have other names for numbers. For instance, an Australian aboriginal population, the Walpiris, combines the terms "one" and "two" to create enough combinations to enumerate sets containing up to 4 items (Dehaene 1997). The same goes for some tribes in the Torres district (Northern Australia), where indigenous people use the words "urapun" and "okosa" to identify "one" and "two", and the expressions "okosa-urapun" (i.e. 2 + 1) and "urapun-urapun" (2 + 2) to designate three and four. Furthermore, they use the word "ras" which means "a lot" (Ifrah 1994). The same features can be found in Africa (the Bushmen of Southern Africa, the Zulu people and the Pygmies of Central Africa), in South America (the Botucodus in Brazil, and the indigenous population of the Tierra del Fuego), and in Sri Lanka (the Veddas). Even if a base 2 numeral system allows for the creation of infinite numbers, it is clear that this procedure entails problems when big quantities are involved. Yet, ancient objects that have been found and date back to more than 20,000 years ago prove that our ancestors were already able to make accurate calculations with big quantities. Researchers have indeed found bones presenting regular cuts or signs grouped together in unordered sequences. These objects were used as counting support to keep track of the population, keep a record of weapons and cloths, applying the principle of one-to-one relation between two sets containing the same numerosity. Hence the question: did these practices precede or follow the establishment of a rich vocabulary to designate numbers?

Obviously, it is extremely difficult to answer this question if we think about prehistoric times, nevertheless it is possible to obtain interesting insights by taking a look to the studies on existing populations that have a limited numerical vocabulary. The languages spoken by these populations, which often are under the threat of extinction, can offer a rare opportunity to establish the width and limitations of non-verbal arithmetical skills. Since their earliest missions on African soil, Western observers have collected several information and data on the presence or absence of numbers in groups belonging to populations seen as primitive and, therefore, more similar to ancient human communities. On the basis of these fragmented observations, made primarily by administrators, missionaries, and traders, ethnographers have developed partial inquiries that focused on wider groups or tribal groups sharing a language or living in the same region, but did not include fieldwork. These comparative data, processed mainly on a second-hand basis, have nevertheless played an important role in the formulation of general anthropological theories on the evolution and generalisation of human societies. These theories were based on the opposition between primitive populations and developed societies, on the opposition between a logical mindset and a pre-logical mindset, on a rudimentary processing or a non-processing of numbers by "primitive" populations (Squillacciotti 1996), establishing in this way a "ranking" of human communities where those at the bottom were those who mastered only a strongly limited amount of numerals (i.e. only one), which entailed a limited capacity and, virtually, a total lack of interest for quantification and enumeration (Tylor 1871; Levy-Bruhl 1912). Even Karl Menninger's abridgement on the cultural history of numbers, despite its encyclopaedic value, is entirely based on this primitivist hypothesis of an alleged prelogical mindset (Menninger 1957). Menninger's work was later replaced by the monumental encyclopaedia on the universal history of numbers by Giorgio Ifrah (1994). Yet, if we consider the issues tackled, even Ifrah's work—which has been translated into several languages—is mainly based on the work by Levy-Bruhl on mental functions in primitive societies (Levy-Bruhl 1912). What stands at the core of this primitivism that is still so in vogue? Firstly, the most basic or simplest systems have a direct relation with nature. Ifrah perfectly illustrates this concept when he contends that "primitive" populations consider numbers as something that can be felt, perceived, learnt, as much as a smell, colour, noise or the presence of another individual or item belonging to the external world can be felt. In other words, these "savages" pay attention only to the changes in the environment within their field of vision, according to a direct relation subject-object. Their ability to understand abstract numbers is therefore limited to what their natural skills allow them to recognise "at first glance" (Ifrah 1994, p. 29). This fact would explain why several populations stopped at the famous sequence: "one, two, three, many". Starting from this, evolution would take place following three stages, with the first stage allowing to clarify the confusion created by a numerous and undetermined quantity, thanks to the use of a certain number of practical methods based on the one-to-one relationship (couples of stones, small sticks, fingers or body parts). The second stage is marked by the use of names of body parts to enumerate concepts that are "insensitively half-abstract and half-concrete", without reaching a full-fledged status of "names for numbers". Then, in the third stage, once the name of the number has been created and adopted, it becomes as useful as the item it represented in the beginning, until eventually the relationship between the two elements completely disappears. One of the evidence that highlight this concrete idea (which incidentally represents a reference point in several school relations and expositions) is the fact that professional anthropologists have long ago lost interest in the enumeration systems, the learning of number chain, the representation of numbers, and the calculation procedures in their own right.

An exception is represented by the studies by some "Africanists" on the symbolism of numbers in West Africa that account for the privileged place that some numbers enjoy in the cosmology of these populations (Nicolas 1968) or the sense and function of numbers associated to males and females (Fainzang 1985). Not even these authors, though, prove the ubiquity of numerical systems in these groups, despite emphasizing the coherence that the systematic combinations of these peoples have with the representations of individuals on the one side, and the social spaces that define them on the other.

In North Region of Cameroon, for example, children are assigned a name-number in relation to their birth and sex even before receiving nicknames (Collard 1973). The different combinations and replacement possibilities that anomalies and accident in life can entail (a change of the order because of the premature death of an older sibling and the possible inversion of the correspondence between the order of birth and order of age) represent specialised manipulations on the order of numbers which, most likely, tend to be disregarded by researchers focussing only on the formal enumeration of numbers in societies. Some researchers took an interest in the question of currency in these societies. Examples are the study by Robert Gray (1960) on the "goat currency" of the Sonko people, the studies by Maurice Godelier (1969) on the "salt currency" in the Bornja people, the analysis by Michel Panoff (1980) on precious objects and payment means in Oceania. These studies had the common goal of assessing the concept of currency in these populations, launching a debate on the conditions of circulation that became so useful in tackling the concrete-abstract question related to numerical systems in societies lacking written traditions.

From this perspective, Daniel De Coppet (1968) carried out interesting research in the Solomon Islands, showing that the local population did not measure surfaces such as those of forests, sky, and sea. On the contrary, this population measured vertical and horizontal lengths using a unit called "brasse" (breaststroke), identified by the distance going from the extremity of the right thumb to the extremity of the left thumb. This unit was used to measure items such as houses, gardens, fishing nets, etc., but also pearl necklaces, a tenured currency in the country and therefore "brasse" units are used for exchanges in the most important events of the daily lives of individuals, tribes, and villages. One "brasse" currency unit is divided into twenty-four lengths, from right to left, that borrow the names of anatomical parts or ornaments that adorn the men's arms. Therefore, the ability of this population to manipulate numbers and the richness of their numerical system cannot be ignored. If we limited our inquiry to a simple verbal enumeration of a succession of names of numbers that would reach only up to "eight" and would be translated as "many", we would not notice that we are dealing with a specific indigenous mindset related to totality and multiplicity, and that the stress given to the first 9 numbers—due to their symbolic value—does not prevent them from manipulating sequences represented by the 50 pearls of their currency or the group of 24 subdivisions that make up a concrete measurement unit.

Jadran Mimica (1988) performed a similar study on the Iqwayas, a tribe living in Papua New Guinea. Iqwayas enumerate numbers orally while representing them figuratively and gesturally. Only the first four numbers are identified by specific terms, but it is possible to count up to big numbers with the use of fingers and toes. The system works on a 20 base (represented by all fingers and toes), with an auxiliary 5 base (represented by the fingers of one hand). All fingers and toes represent 20 and are equal to one "man". At the same time, each finger can also identify an individual and, therefore, each finger may represent 20, allowing the entire human body to represent the number 400. This implies that big numbers can be counted following a calculation process that works thanks to the "use of one finger to represent a set of fingers". Furthermore, it implies also that the physical presence of another individual is not necessary for each multiplication. Finally, it implies also that counting is an abstract process, even despite the fact that for the individual counting on his fingers it represents a physical process with a figurative-gestural nature, because for each new exponentiation of the base, he refers to a virtual or absent new individual.

In addition to this, other populations have developed an ingenious procedure to evaluate and memorise big quantities without using specific words and objects. The island inhabitants of the district of Torres (Dehaene 1997), for example, can count up to 33 items by indicating in accurate order the different parts of their body: they start with the pinkie on their right hand (1) and move towards the thumb (5), then they move to the wrist (6), elbow (7), shoulders (8) and torso (9). At this point, they move to the left arm, following the same order. Once they have reached the pinkie on their left hand (17), they move with a similar trajectory from the small toe on their left foot (18) to the small toe on their right foot (33) going through their ankles (23, 28), knees (24, 27) and hips (25, 26). Obviously, the names of the body parts do not represent a numerical code, but they are used as formulations to translate quantities. The first rudimentary words to indicate numbers, arbitrary symbols designating quantities, have certainly appeared as an abstraction of these procedures. Each quantity was unambiguously associated to a body part and by uttering the name of this body part it was possible to identify a quantity. "Tail" or "shoulders" obtained a numerical meaning (7 and 8) and evoked immediately the quantities associated to them, without referring to the gestures one had to do to point at them. Therefore, they obtained an abstract meaning, despite the fact that the initial formulation related to the origin of their meaning and it was concrete, since it recalled an actual body part.

These studies testify therefore the existence of more or less advanced numerical systems even in societies considered as "primitive". In some cases, theoretical repercussions are exaggerated. An example of this is provided by Mimica, who broadens his monograph by addressing the concept of infinity encountered in the culture at the centre of his studies, even putting forward the idea of comparing it to Cantor's theory on transfinite ordinal numbers.

3.2.1 The Mundurukú People

To broaden the studies on the existing relationship between language and arithmetic, Pierre Pica, Cathy Lemer, Veronique Izard and Stanislas Dehaene (Pica et al. 2004) decided to study the Mundurukú people, a population living in an autonomous territory of the Para state in Brazil, whose language belongs to the Tupi family and has names for identifying the numbers from 1 to 5. Some Mundurukú are more or less able to speak Portuguese and children are educated up to an age of 10 years in their mother tongue, while from 10 to 14 years their education takes place in Portuguese.

However, not all children attend school, and the lessons are often very rudimentary and taught by young professors belonging to the local community. Therefore, the level of bilingualism is uneven: in general, elderly people, women and children are monolingual, while adults have some knowledge of Portuguese. Similarly, as far as numbers are concerned, several locals know the sequences of numbers in Portuguese even if they never use them in their daily lives. The research team performed a series of quite sophisticated tests on 55 people belonging to the Mundurukú population.

The first test aimed at testing the level of numerical vocabulary knowledge. The subjects were divided into different groups, according to their level of exposure to numerical systems different from the monolingual indigenous system. Two groups of strictly monolingual and uneducated adults and children were compared to groups of adults and children partially bilingual and educated. Furthermore, contrary to the experiments performed by Everett and Gordon, a control group was established by involving French adults. In the test, Mundurukú were presented with stimuli containing 1 to 15 dots, casually scattered over a surface, and asked to indicate how many dots they could see. In order to make sure that participants were responding on the basis of the numerosity presented to them and not on other low-level variables, each numerosity made of 1 to 15 dots was presented twice, performing different controls on non-numerical parameters. In the first series of tests, the extensive variables (total lightness and total occupied space) were equal for all numerosities, while in the second series of tests, intensive variables were levelled (format of the dots and space among them). In both situations, no variations were recorded, with the exception of the lack of a word to identify "five" among young people. Over 5, it was usual to hear expressions such as a little (*adesu*), a lot (*ade*), or a certain quantity (*buru maku*), all using approximate quantification labels.

Furthermore, Mundurukú used several other expressions such as "more than one hand", "two hands", "some big toes", and also long labels such as "all fingers on one hand" or "even some more" (to provide an answer to the case of 13 dots). The use of a succession of numbers to enumerate accurate quantities did not represent a familiar action for Mundurukú. Even if some individuals, when explicitly asked to count the objects presented, strived to use a numerical sequence by using their fingers and toes, the majority of them spoke out numbers without counting. These data confirm the fact that Mundurukú select their oral answers on the basis of a knowledge of an approximate number rather than an accurate counting procedure. With the exception of 1 and 2 dots, all numerals are used in relation with a wide range of approximate quantities rather than an accurate number. For example, the word used to identify five, which can be translated as one "hand" or one "fist", was used to identify 5 as well as 6, 7, 8, or 9 dots. Therefore, if it is true that the concepts related to numbers are visible only when numerals are available, it would be fair to expect that Mundurukú have strong difficulties in processing big numbers (Fig. 3.1).

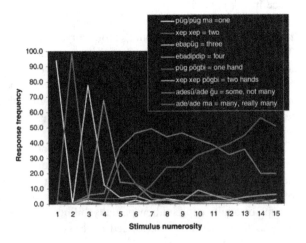

Fig. 3.1 How Mundurukú call numerosities (Pica et al. 2004)

Researchers have tested this conclusion with two targeted tests: one required Mundurukú to compare sets of dots, while the other asked them to add and compare sets of dots. In the comparison tests, subjects were simultaneously presented with two sets placed one next to the other: the one on the left was black, the one on the right was red. They were then asked which set contained the biggest number of dots. It must be noticed that 50% of times, the biggest set was placed on the left, while in the other cases it was placed on the right. In addition to this, the ratio between the two numerosities was changed: as we know, this ratio determines the level of difficulty of the comparison (R = 1,2; 1,3; 1,5; or 2,0). For each R-value, three couples of numerosities of different formats were used (small: 20–30 dots; medium: 30–60 dots; big: 40–80 dots). The answers provided by the test subjects (70.5% of correct answers) did not show substantial discrepancies among the groups. This led to the conclusion that the level of bilingualism and education of subjects did not affect their level of performance.

In the control group composed of Western people, the numerical comparison performances were affected by the distance effect: performances improved according to the ratio between the numerosities under scrutiny, both when the stimuli were presented as sets of items or as symbols with Arabic numerals. The same effect was observed among Mundurukú: their performances deteriorated when the ratio between the two numerosities under comparison went down from 2 to 1.5, 1.3, or 1.2. Reaction times were also affected by the distance effect: they were faster when two distant numerosities were involved rather than in the case of two near numerosities. The value obtained by Mundurukú was 0.17, only a little bit slower compared to the 0.12 recorded by the control group. Subsequently, the researchers tested the Mundurukú' abilities in approximate operations with big numbers by assigning them an approximate addition task.

At the beginning of this test, an empty jar was shown to the subjects. The jar was presented in vertical position, suggesting that the content would not have fallen. Two sets of dots fell into the jar from the upper part of the screen. The two sets of dots never appeared simultaneously but always followed each other (each set moved for a time of 5 s). Then, a third set of dots appeared on the right side of the jar. At the end of the entire sequence, the participants had to decide if there were more dots inside or outside the jar (Fig. 3.2).

To keep under control the non-numerical parameters, the researchers resorted to the same procedures used in comparison tasks: the ratio between the sets n1 + n2 and n3 were changed. In the first three tasks, the ratio was 4, while in successive tests ratios of 1.3, 1.5 and 2.0 were presented (16 tasks for each ratio, numerosities between 30 and 80, and the two parts of the addition always had a ratio of 2:1, 1:1 o 1:2). All groups, including those of monolingual adults and children, recorded good performances (80.7% of correct answers), both for the sets used in tests with intensive and extensive parameters (89.5 and 81.80% of correct answers respectively). Even if the performance level was different between the two series of stimuli, it must be noticed that the same differences recorded among Mundurukú were recorded also in the control group composed of Western people (Fig. 3.3).

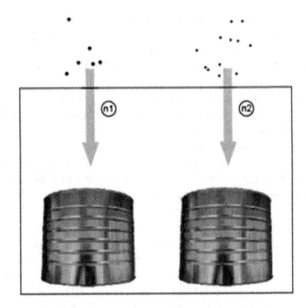

Fig. 3.2 Representation of the experimental setting for comparison tasks: two sets of dots fall into empty jars

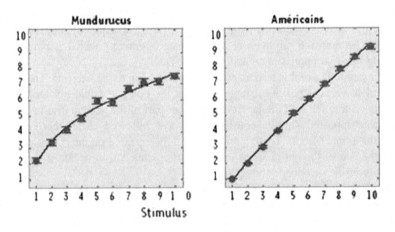

Fig. 3.3 Answers provided by Mundurukú and the control group (Izard et al. 2008)

Mundurukú therefore did not seem to have any difficulty in adding and comparing approximate quantities. On the contrary, they seemed able to provide the same level of accuracy of the Western control group. Furthermore, they were also able to make mental representations of big numbers (up to 80), numbers well above those that they could enumerate. In addition, they did not mix up numbers with other variables such as the format or density of dots. They spontaneously implemented the concept of addition, subtraction, and comparison to the less distant representations. The same was recorded in monolingual adults and small children that had never learnt formal arithmetic, showing that an advanced numerical competence (if only in an approximate form) can exist without a well-developed numerical vocabulary.

In a later experiment, the same researchers evaluated the hypothesis that Mundurukú could behave as the Western children involved in Feigenson's research (Feigenson et al. 2004). As we have explained in the previous chapter, the data collected by Feigenson in 10- and 12-month-old children showed that approximate numerical representations could not be compared among them nor be translated from one format to the other. In general, in Feigenson's experiment, whenever one of the two sets contained more than 4 elements, children failed at their task. These results led to the conclusion that children could make an analogical representation of quantities when objects appeared, but they were not able to make the same representations when there were too many objects involved. The author contended that the acquisition of a numerical vocabulary or the practice of counting became at that point a paramount factor to perform this kind of manipulation.

As in the case of children, also Mundurukú do not practice counting, neither in their mother tongue nor in Portuguese. Therefore, also in their case, when elements are added one by one and the total number of objects is hidden to them, they should not be able to access a numerosity that goes beyond four elements. To test this hypothesis, Pica and colleagues designed a comparison test in which two sets were hidden. At the beginning of each test, the researchers showed the subjects two empty jars. Then, a first set of dots fell in the jar on the left. Subsequently, the jar disappeared and a second set of dots fell in the jar on the right. In order to prevent subjects from using non-numerical parameters to formulate their answers, two series of stimuli were created, following the experimental procedures of the previous experiments. In the first series, the intensive parameters were the same between all numerosities, while in the second series the extensive parameters were levelled. The dots appeared both simultaneously (simultaneous presentation) and falling one after the other in the jar (sequential presentation), using couples of numerosities in a 1:2 ratio. As far as sequential tasks were concerned, the falling frequency of the dots in the jar was stable and the total time of presentation increased, as the numerosity got bigger.

The data collected referred to 47 Mundurukú subjects divided in 5 groups: monolingual adults unable to count in Portuguese (8), monolingual adults able to count in Portuguese (16), bilingual adults able to count in Portuguese (6), children unable to count in Portuguese (7), and children able to count in Portuguese (10) The results highlighted a higher level of performance when the numerosities were distant as far as their ratio was concerned, a fact recorded among all the groups. The participants that could count in Portuguese, though, had better performances than those who did not, in particular in the sequential tasks (probably due to the fact that they counted the dots while falling). The most significant data for our inquiry refer to the group of participants that could not count (8 adults and 7 children): if Mundurukú behaved as Feigenson's children, they should have failed in sequential tasks. On the other hand, had they used indifferentially their approximate number sense to solve these tasks, the results in both cases should have been the same level of performance. The analysis made by Pica shows an effect on the type of performance (correct answers rate: 78% for the stimuli presented simultaneously, 68% for those presented sequentially).

These results suggest then that Mundurukú and children use different representations according to the fact that sets are presented sequentially or simultaneously. Yet, contrary to Feigenson's children that answered randomly in the case of big sets presented sequentially, Mundurukú provided good answers for all numerosities in all types of presentation. This suggests that there is a different effect at play in the two cases: in children, the effect leads to an inability to go from a representation in the form of an attention indicator to an analogical representation, while in Mundurukú it seems more likely that the observed effect reflects different accuracy levels of numerical representations for sets presented simultaneously or sequentially.

Confirming this hypothesis would require a study on a control group composed of Western subjects, to assess whether they show substantial difficulties when sets are presented sequentially. Nevertheless, it is likely that the accuracy of numerical representations is affected by the pace of the sequential presentation: when elements are presented very fast, it can be assumed that subjects use a mechanism similar to the accumulator described by Charles Gallistel.

On the contrary, when elements are presented one by one and at slow pace, this mechanism proves useless and successive additions become necessary: the estimated total will then be less accurate than the one provided by the accumulator at a faster presentation pace. The hypothesis of a "sense of approximate number" contends that when oral or written symbols are not available, a number is represented only approximately, with an internal uncertainty that increases along with the increase in the number (Weber's law): after 3 or 4, this system can only perceive a negligible difference between a precise number n and the following number $n + 1$. If Mundurukú, as it seems, have this non-verbal and inaccurate way of representing numbers, they should fail in all tasks requiring the processing of precise numbers.

To test this last hypothesis, researchers used a task of precise subtraction. At the beginning of the test, an empty jar was shown. Then, some dots fell from the upper part of the screen into the jar. After a brief period of time, some other dots fell from the jar, disappearing from the screen (movement duration for each set: 2 s, time between the two movements: 6 s). The numerosity presented was between 1 and 8 and the result between 0 and 2. The Mundurukú subjects had to answer by choosing the correct result among three alternatives (0, 1, or 2 remaining points). In a second version of this test, where an oral answer was requested, participants had to verbally describe the content of the jar at the end of the video (it must be noticed that zero does not exist for Mundurukú, but the participants spontaneously said sentences such as "nothing has left"). The results showed that the performance level was slightly better in the bilingual and educated groups (particularly when more than 5 dots were shown). At the same time, all Mundurukú groups' performance level was clearly worse compared to the one recorded by the Western control group, whose performance was not minimally affected by the number format. Nevertheless, the failure of Mundurukú in this exact subtraction task was not due to a low level of education, since Mundurukú had great performances (almost perfect) when the numerosities presented were under 4. Their success with smaller numbers reflects the use of a preverbal code or a parallel identification system of objects similar to the one used by children and apes. Concretely, in the exact task presented, Mundurukú still used approximate representations subject to Weber's law, while the Western control group solved it by using exact calculations.

In a nutshell, Western cultures have put forward the idea that new concepts arise when children learn to use numerical vocabulary. Nevertheless, by simply studying children, it is hard to evaluate the limits of the role of cultural learning (in particular, the importance of a gradual familiarisation with the names of numbers) and cerebral maturation, regardless of the surrounding environment. The studies on Mundurukú make a clear distinction among these factors. Mundurukú are able to access numerosities by going beyond their vocabulary constraints and are aware of the fundamental laws that govern the evolution of the cardinal features of sets (union and addition, separation and subtraction, order of numerosities). In approximate arithmetical tasks, when they were asked to roughly estimate the results of operations, the behaviour of indigenous people was similar to those of Western adults. On the contrary, when asked to perform arithmetical tasks that required an exact result, they kept providing approximate answers.

At the same time, they are able to evaluate the exact difference between two sets of objects when the circumstances allow them to verify if the elements of these sets have a biunivocal correspondence. They do not lack the abstract concept of perfect equality, even if they cannot always apply this concept in specific contexts. Mundurukú own an approximate number sense and use this knowledge on a wide range of numerosities. As for tasks involving big numerosities, the behaviour of this population is affected by the same distance effect recorded in Western adults, children, and animals.

These conclusions are in line with the projection of the representation model of numerosities on an internal *continuum*, both in the case of the comparison of numerosities, their addition, or subtraction. The internal Weber's fraction w, the main parameter in the model that measures the level of inaccuracy of numerical representations, is around 0.17 among Mundurukú in the comparison tests, similar to the usual result recorded in Western subjects (in Pica's study $w = 0.12$). All these results therefore provide new evidence that corroborate the hypothesis linked to the existence of a "number sense", a language-independent skill universally shared among humans of all societies, children, and several animal species.

The fact that Mundurukú can compare big quantities presented sequentially points to the fact that they use the same representation system of approximate numerosities without considering the format of the numbers involved. The observations on Mundurukú, who have a limited vocabulary but own all the cognitive systems needed to develop a language, suggest that Chomsky's hypothesis that there is a bond between numerical competences and language requires at least a couple of adjustments (Chomsky 1957). First of all, it must be admitted that there is a non-verbal representation of near numbers and a genuinely conceptual competence for learning and manipulating approximate numerical quantities. This competence seems to be language-independent, since it has been found in Mundurukú, new-borns, and several animal species. Its existence was not explicitly recognised by Chomsky, who showed little interest in the origins of arithmetical skills in humans. Nevertheless, the hypothesis of a "number sense" (a set of cerebral wires that allow humans to understand the cardinal features of a set of objects and the fact that these features are not affected or do not change according to the operations applied to this set) is in line with the explanation model developed by Chomsky, which states that human cognitive skills are based on specialised inborn systems. The studies presented above also suggest that the ability to manipulate exact big numbers is characteristic of few cultures, such as ours, which have at their disposal a wide vocabulary for exact numbers and syntactical rules to combine them and create an infinity of number names. This aspect is compatible with Chomsky's hypothesis that states that the combination features of language play a pivotal role in the establishment of arithmetic.

Nevertheless, the bond between language and exact calculation is certainly less important than what Chomsky suggests. Without counting, which requires a fast enumeration of a succession of numbers, it is impossible to successfully complete the majority of exact arithmetical tasks. If Mundurukú are allowed to use alternative strategies to evaluate the exact equality of two sets of objects (e.g. let them verify the correspondence of the elements of the two sets one by one), they can distinguish even big numerosities that differ only by one unit.

The effect of language in the exact subtraction task that we presented is therefore related to a difference in performance—to use another differentiation made by Chomsky—in other words, the factors that determine the ability to succeed in the task rather than a genuine difference in conceptual skills. The successful resolution of the task does not depend only on the control of the concept of exact number (an abstract arithmetical competence that Mundurukú showed to own), but also on other factors that Chomsky would categorise as "external", such as the ability to count effectively.

Finally, the results achieved contradict Gordon's conclusion that supported Whorf's hypothesis and the idea of a new linguistic determinism by saying that numerical concepts in the Pirahã population were "incomparable" to ours (Gordon 2004). Starting from the assumption that the individuals belonging to the Pirahã population and interviewed by Gordon were equipped with the same conceptual heritage of Mundurukú, we can postulate that these individuals failed in the numerosity task presented by the author because he failed in developing an experimental setting suited to testing the real skills of the subjects involved in his experiment. In the light of what we currently know and of the low number of Pirahã participants in the tests performed by Gordon, his hypothesis seems irrational and too simple as much as the one contending that their numerical cognition is radically different from ours.

3.3 Western and Eastern Societies

Establishing a possible comparison between West and East using these specific labels is a questionable exercise: without more clarity on the meaning of the term "West", the word may become misleading, since it literally identifies the entire Western hemisphere and hence a wide geographical area inhabited by different ethnic groups. The same goes for the term "East". At a glance, the complex Western cultural system can be boiled down to its double cultural matrix: the first one is represented by Greek philosophy, the mother of Western philosophy *par excellence*, guided by the trio composed by Socrates/Plato/Aristotle; while the second matrix is rooted in the Judaeo-Christian religion. The constructive combination of Aristotelian philosophy and Christian religion, which is strongly focussed on individualism, left a clear mark on Western civilisation, defining a mentality that focus on the idea of the individual, both as part of the society and independent subject, in a sort of "insulated perfectionism" harmonised to Aristotle's "Golden Mean". This idea stands in stark opposition to the Eastern mindset, which focuses more on social organisation rather than individuals.

China, the "radiant sun" that shines on the civilisation of South-East Asia, has at the core of its culture the *Tao* and the *ying-yang* complementarity, which combined with Confucianism and Buddhism defines its cultural DNA. In the Eastern world, cardinal virtues are summed up by the concepts of *ren* and *li*. It is difficult to provide a precise translation of these terms: *ren* can be translated as benevolence, kindness, selfless human love; *li* represents honesty and a proper behaviour. In others words, *li* shapes the interpersonal channels through which *ren* flows from one individual to the other. In Chinese philosophy there is no clear division between black and white, not even in the interpenetrating colours of the *ying-yang* circle that symbolise the *Tao*. The focal point of the *Tao* emphasizes that contrasting couples are not in opposition and do not mutually exclude each other, but rather complement and mutually include the other part. Everything can be interpreted according to different shades by following multiform and non-Aristotelian logics.

On the one side, the Aristotelian bivalent logics pinpoints the way Western cultures conceive the world (Euclidean geometry, the first structured language in the history of mathematics, is a genuine embodiment of the Aristotelian bivalent logics model), while on the other, the foundations of reality are based on the Confucian methods of transmission of *Tao* and the book *I Ching* ("Classic of changes", the fundamental text to understand Chinese metaphysics and the reasoning schemes that were developed over the years in the history of Chinese mathematics).

As far as mathematical language is concerned, researchers have always been fascinated by a very peculiar characteristic of the Chinese language, which is how words are built together and how this system is perfectly compatible with counting. This system is often named as a surprising example that showcases the way in which culture can positively affect numerical cognition (Butterworth 1999).

In Chinese, the names corresponding to the first nine numbers *(yi, èr, san, si, wu, liù, qi, ba, jiu)* are combined with multipliers such as 10 *(shi)*, 100 *(bai)*, 1000 (*qian*) and 10000 (*wàn*) to build numbers, following a rigorous 10 base breakdown-rule. According to these rules, 327 is pronounced *san bai èr shi qi*.

More concretely, in Chinese (and in other East Asian languages such as Japanese, derived from Old Chinese), the words used for calculations strictly follow the decimal logic, contrary to what happens in European languages such as French, English, or Italian. Children learning to count in France or Great Britain find themselves in the position of deciphering "unknown" numerical terms such as "eleven", "twelve", or "twenty". They are "unknown" because the logical formation of these words is not so evident, since there is no evidence that "eleven" means "ten plus one". In Chinese, on the contrary, "eleven" means "ten-one", "twelve" means "ten-two" (*shi-er*), "twenty" means "two-ten" (*er-shi*), etc. These terms are therefore easily understandable and learnable, since they use numerals already known and follow to the letter the decimal logic on which the numerical position notation system is based. David Geary summed it up by saying that thanks to the coherence of numbers in Eastern Asia, children tend to make fewer calculation mistakes, learn the concepts of calculation and number at an early age, make fewer mistakes in solving arithmetical problems, and understand basic arithmetical concepts—such as those used in trade—much earlier than their American or European counterparts (Geary 1994, p. 244).

Another feature of the Chinese language that is deemed to have an important impact of numerical cognition is the relative speed of pronunciation of Chinese numerals, since it seems to increase the impact of numbers on memory, strengthening the mental calculation skills (Geary et al. 1993; Chen and Stevenson 1988; Stigler 1984). Then, there is the important fact that Chinese is a tonal language (in Mandarin, the five existing tones can basically be described as 1: constant, 2: rising, 3: dipping tone, alternatively falling and rising, 4: falling, and finally 5: neutral). This means that all words—numbers included—are sung more than they are spoken and tones help in determining the meaning. Chinese words used to identify the numbers from one to ten correspond to different tones that help in distinguishing them: one = first tone, two = fourth tone, three = first tone, four = fourth tone, five = third tone, six = fourth tone, etc. When uttered one after the other, they create melodic patterns that help memorisation.

It is worth noting that even as far as ordinal numbers in English and French are concerned, children must learn a completely new set of terms such as "first", "second", "third", etc., before they can determine which place is occupied by an item in a set of objects. In Chinese, on the contrary, the formation of ordinal adjectives is extremely simple, since they are created always in the same way and through a straightforward logical process: one simply adds the prefix *di* in front of the number. "The first one" reads *di-yi* (*yi* meaning "one"), "the second one" reads *di-er* (*er* meaning "two") and so on. The Chinese language therefore does not complicate excessively a concept that, at the end of the day, is quite simple.

Furthermore, Chinese is a language in which measurement terms are systematically used when enumerating or identifying objects. In Italian, people use measurement terms only sporadically, for example when people say "cup" in the expression "a cup of coffee". On the contrary, in Chinese, it is impossible to avoid using measurement terms. For example, if one wishes to say "two roads" in

Chinese, it is necessary to add a measurement term: "two + measurement term + roads" (*liang tiao lu*). In this example, the measurement term means "long and sinuous". In Chinese, one cannot say "five books", a measurement term must be placed between the number and the name: "five + measurement term + books" (*wu ben shu*), which means something similar to five publications of books. This linguistic feature leads to three observations. First of all, children learn measurement terms while starting counting items in the real world; measurement terms are therefore an important element in the development of the ability of enumerating items in the real world, even if they might seem of secondary importance. Furthermore, it is possible to assume that the possibility to omit measurement terms —making enumeration a little bit harder—emphasizes the fact that people tirelessly count things. Finally, they highlight some characteristics of the enumerated elements, moving the attention to the "length" of roads, the "flatness" of boards, or the "honesty" of people. This could seem completely unrelated to numbers as entities. Nevertheless, since the distinctive features of real-world objects are directly embedded in the enumeration process—through the trio number + measurement term + object—measurement terms help in highlighting the meaning of numbers in a specific context. Finally, it must be noticed that also in its written form, Chinese is intrinsically numerical. For example, the character identifying a "rock", *shi*, is made up of five traits that must be drawn in a strict order: one, two, three, four, five.

Even Chinese vocabularies are organised on a numerical basis, according to the number of traits that make up a specific character or radical (radicals are the basic elements of characters). When looking for a character, it is necessary to count the number of its traits. A similar reasoning can be followed for classic mathematical concepts such as variables, the key concept of algebra and generalisation processes (another key element of algebra). The concepts of "unknown numbers", variables and more generally that of parameter are, broadly speaking, quite difficult to understand. Yet, by analysing the *Jiuzhang Suanshu* text, which is considered the reference ancient text in Chinese algebra, it is clear that since ancient times Chinese ignored the necessity of deductive argumentation and their mathematics was not based on the severity of a well-organised logical technique such as among the Greeks, but rather on the practical use of mathematics by citizens in economic and agricultural contexts. The result was that mathematics was presented as a group of terms developed by following a spontaneous and natural "technical grammar" rather than as harmonic set of knowledge. In the *Jiuzhang Suanshu*, solution procedures (*shu*) follow a procedural structure that exploits the data presented in the text of the problem and express specific quantities and numerical values. For example, in the problem I.9:ì, it is assumed that someone has the quantities 1/2; 2/3, 3/4, 4/5. The question is: how much can you obtain by putting all together? The distinctive elements that characterise the resolution methods proposed are mainly inferred from the textual object represented by a simple list of operations that make up the entire procedure (Kline 1972), where known terms such as "numerator", "denominator", "multiply…" can be found (Chemla 2007). Therefore, even if the resolution procedures are basically arithmetical and numerical in nature, the text clearly emphasizes the search for general algorithms that can be applied in different problems that partially refer to concrete and specific situations. The use of arith-metical operations (multiply, divide…) and the use of terms such as "the multiplier to breakdown", "simplify to reunite", "homogenise", "equal to establish a com-munication", show the demonstrative value of the resolution methods of ancient China, which represent strategic objectives in the search for constants in different calculation procedures: from this perspective, expressions such as "make them equal" and "make them homogenous" can provide concrete indications of algebraic manipulation. To sum up, even if some structural characteristics of Chinese have not been yet object of thorough inquiries, they can affect significantly the devel-opment of numerical cognition.

Even if numerical cognition is not as language-dependent as previously thought by Piaget, it seems to be accessible under linguistic form at least in its "exact" part. In this case, the differences between languages and, broadly speaking, cultures play a direct and undeniable role.

3.4   The Role of Language

Children learn to say "one", "two", "three", etc. thanks to the act of counting, which allows them to label each object. Children hence learn the sequence of natural num-bers as a nursery rhyme that mechanically communicates the passing of time: 3 is the number that comes before 4 and follows 2. Nevertheless, children need much more time to understand that the last word of this temporal sequence may have another meaning, i.e. the cardinal number that indicates how many objects there are in a set. The majority of children cannot understand this principle before the age of 3. When one asks them "how many objects are there?", they systematically answer with a different number than the one they have just enumerated by counting. Borrowing the observation made by psychologist Karen Fuson "the situation described by numbers is not evident at all" (Fuson 1988, p. 3). At the age of 3, children start recognising the numerosities 1, 2 and 3; when they understand the numerosity 4, they find out how the entire list of whole numbers is structured and de facto learn how to recognise all numerosities. In the end, it is their linguistic competence that allows them to go beyond the original non-verbal representations. Without language, many strategies would be impossible, such as those requiring the use of verbal memory, in particular counting and mental calculation. Therefore, oral or written enumeration mobilises a linguistic-verbal system that involves a series of vocabulary and combination rules. Numerical systems are conventional systems that rest on two major principles: (1) lexicalisation, an elementary process that associates a cardinal number to a name; (2) the use of syntax and combination rules (addition and multiplication), which allows for the creation of an infinite number of complex denominations that do not correspond to specific cardinal numbers. Despite some irrelevant local divergence, several languages use a 10 base system in their numerical systems, where each whole natural number has a name and, mutually, the names of numbers are unambiguous. Some authors, led by Noam Chomsky, have identified recursion as the fundamental component driving the development of numerical systems (Hauser et al. 2002).

Recursion is a procedure that recalls a structure established starting from a structure of the same type: in the spoken language, examples of sentences with recursive structure are those containing linked subordinates (e.g. the book that Pietro [that is my friend] gave to me). From a syntactical perspective, language is *recursive* (an example is the following rule on nominal groups that describes how to add an adjective: NOMINAL GROUP = ADJECTIVE + NOMINAL GROUP, where the addition can be made an *x* number of times, since the result will always be a valid nominal group) and the same goes for natural numbers (a whole number N is associated with N + 1). In mathematics, some functions are built in recursive manner, such as successions where the term *Un* is identified by taking a previous term *U n-1*. Among recursive structures, there is a category of simpler structures, which are those obtained through repetitions. Repeating a procedure means applying it more times successively. Natural whole numbers are an example of structures obtained through repetition: all numbers can be obtained by adding one unit to the previous one. In some languages, the name of numbers is constructed on the principle of recursion and the name of big numbers contains often the name of smaller numbers. The most striking example is provided by Asian languages (as we have seen in the previous paragraph, in Chinese all numbers are named following always the same rule). The invention that increased the efficiency of numerical notation was the establishment of the principle of position value. A system of numerical notation follows the principle of position value when the quantity represented by a numeral varies according to its position in a number. Three identical numerals therefore can refer to three different sizes in the number 222: two hundreds, two tens, and two units. In a notation system with position value, there is always a privileged number called base. The concept of base was developed already by Babylonians and it establishes an efficient combination rule. In other terms, the base is the number from which names are repeated following rules that govern their combination. On the side of the 10-base system, times and angles are measured in a 60-base system. Also 20-base systems, used in ancient times by the Mayas and Aztecs, have survived in some counting systems and languages, such as Danish, French and German, where some numbers are expressed as multipliers of twenty (e.g. *quatre-vingts*, i.e. "four times twenty", which means 80 in French). Therefore, regardless of the base, in almost all languages, basic numerical words can be combined following the principle of recursion and accurate syntactical rules, in order to precisely and non-ambiguously build all thinkable numbers.

References

Berlin, B., & Kay, P. (1969). *Basic color terms: Their universality and evolution*. Berkeley: University of California Press.
Butterworth, B. (1999). *The mathematical brain*. London: Papermac/Macmillan.
Chemla, K. (2007). *Matematica e cultura nella Cina antica*. Torino: Einaudi.
Chen, C., & Stevenson, H. (1988). Cross-linguistic differences in digit span of preschool children. *Journal of Experimental Child Psychology, 46,* 150–158.
Chomsky, N. (1957). *Syntactic structures*. The Hague: Mouton.
Collard, C. (1973). Les 'noms-numéros' chez les Guidar. *L'Homme, XIII*(3), 45–59.

Davies, I. R. L., & Corbett, G. G. (1997). A cross-cultural study of colour grouping: Evidence for weak linguistic relativity. *British Journal of Psychology, 88*(3), 493–517.

De Coppet, D. (1968). Pour une étude des échanges cérémoniels en Mélanésie. *L'Homme, VIII*(4), 45–57.

Dehaene, S. (1997). *The number sense*. New York, Cambridge (UK): Oxford University Press, Penguin Press.

Everett, D. (2005). Cultural constraints on grammar and cognition in Pirahã: Another look at the design features of human language. *Current Anthropology, 46*(4), 621–634.

Fainzang, S. (1985). Les Sexes et leurs nombres. Sens et fonction du 3 et du 4 dans une société burkinabé. *L'Homme, XXV*(4), 97–109.

Feigenson, L., Dehaene, S., & Spelke, E. S. (2004). Core systems of number. *Trends in Cognitive Sciences, 8*, 307–314.

Fuson, K. C. (1988). *Children's counting and concepts of number*. Berlin: Springer-Verlag.

Geary, D. C. (1994). *Children's mathematical development: Research and practical applications*. Washington, DC: APA.

Geary, D., Bow-Thomas, C., Fan, L., & Siegler, R. (1993). Even before formal instruction, Chinese children outperform American children in mental arithmetic. *Cognitive Development, 8*, 517–529.

Godelier, M. (1969). La monnaie de sel des Baruyas de Nouvelle-Guinée. *L'Homme, IX-2*, 5–37.

Gordon, P. (2004). Numerical cognition without words: Evidence from amazonia. *Science, 306* (5695), 496–499.

Gray, R. (1960). Sonjo bride price and the question of African wife purchase. *American Anthropologist, 62*, 34–57.

Hauser, M. D., Chomsky, N., & Fitch, W. T. (2002). The faculty of language: What is it, who has it, and how did it evolve? *Science, 298*, 1569–1579.

Ifrah, G. (1994). *Histoire universelle des chiffres*. Paris: Robert Laffon Bouqins.

Izard, V., Pica, P., Spelke, E., & Dehaene, S. (2008). Comment les nombres se répartissent dans l'espace. *Médecine sciences, 24*, 1014–1016.

Kay, P., & Regier, T. (2003). Resolving the question of color naming universals. *PNAS, 15*, 9085–9089.

Kline, M. (1972). *Storia del pensiero matematico*. Torino: Einaudi.

Lave, J. (1988). *Cognition in practice. Mind mathematics and culture in everyday life*. Cambridge: Cambridge University Press.

Levy-Bruhl, L. (1912). *Les fonctions mentales dans les sociétés inférieures*. Paris: Vrin.

Menninger, K. (1957). *Number words and number symbols*. Cambridge: MIT Press.

Mimica, J. (1988). *Intimations of infinity. The cultural meanings of the iqwaye counting and number system*. Oxford: Berg.

Nicolas, G. (1968). Un système numérique symbolique: le quatre, le trois et le sept dans la cosmologie d'une société hausa (vallée de Maradi). *Cahiers d'Etudes africaines, 32*, 566–616.

Nunes, T., Schliemann, A., & Carraher, D. (1993). *Street mathematics and school mathematics*. Cambridge: Cambridge University Press.

Panoff, M. (1980). Objets précieux et moyens de paiement chez les Maenge de Nouvelle-Bretagne. *L'Homme, XX*(2), 5–37.

Pica, P., Lemer, C., Izard, V., & Dehaene, S. (2004). Exact and approximate arithmetic in an amazonian indigene group. *Science, 306*, 499–503.

Pinker, S. (1994). *The language instinct*. London: Penguin Books.

Rosch Heider, E. (1972). Universals in color naming and memory. *Journal of Experimental Psychology, 93*(1), 10–20.

Squillacciotti, M. (1996). *Antropologia del numero*. Brescia: Grafo Edizioni.

Stigler, J. (1984). Mental abacus. The effect of abacus training on Chinese children's mental calculations. *Cognitive Psychology, 16*, 145–176.

Tylor, E. B. (1871). *Primitive culture*. London: Murray.

Uchikawa, K., & Boynton, R. M. (1987). Categorical color perception of Japanese observers: Comparison with that of Americans. *Vision Res, 27*, 1825–1833.

Whorf, B. L. (1956). *Language, thought and reality*. Cambridge, MA: The MIT Press.

Chapter 4
Dissociations Between System 1 and System 2

Abstract Calculation ability represents an extremely complex cognitive process. It has been understood to represent a multifactor skill, including verbal, spatial, memory, and executive function abilities. In this chapter, we will deal with it by calculation disturbances are analyzed. Specifically, evidence from brain-damaged patients indicates that deficits in mathematics can follow injury to either cerebral hemisphere, but that the nature of the impairment will differ depending upon the locus of the cerebral insult.

Keywords Acalculia · Developmental dyscalculia · Turner syndrome Double dissociation

4.1 Calculation Disorders

The loss of the ability to perform calculation tasks due to brain disorders is called "acalculia" or "acquired dyscalculia". Acalculia is often mentioned in neurological studies and in neuropsychological research, but targeted analyses of this disorder are quite rare. Psychological and neuropsychological cognition evaluations tend to include tests of calculation skills, but it is quite hard to find a specific exam targeting acalculia in scientific literature. Historically speaking, before the recent development of cognitive neuropsychology, scientific papers often referred to calculation disorders, even though many of these cases were interpreted as a consequence of language deficits (aphasia). Lewandowskey and Stadelmann, for example, published the first detailed report of a patient suffering from calculation disorders already in 1908 (Lewandowskey and Stadelman 1908). The patient showed strong difficulties in mental and written calculations as a consequence of a brain lesion. The authors identified difficulties in the reading of arithmetical symbols, despite the patient being able to correctly solve arithmetical operations. This paper represents an important point of reference in the development of the concept of acalculia, because for the first time calculation disorders were considered as something different, unrelated to language disorders.

© The Author(s) 2018
M. Graziano, *Dual-Process Theories of Numerical Cognition*,
SpringerBriefs in Philosophy, https://doi.org/10.1007/978-3-319-96797-4_4

The first researcher to use the term acalculia was Salomon Henschen (1925), defining it as a specific deficit independent from other types of disorders, even if this term was used to refer to all kinds of problems related to the use of numbers. Henschen examined 305 clinical cases in academic literature where calculation disorders related to brain damage were reported. Among these cases, he found patients presenting calculation disorders without vocabulary problems, hence corroborating the hypothesis that there was a specialised anatomical sub layer devoted to arithmetical operations that was separated—even if located in proximity—to the areas devoted to language and musical skills (he pointed to the third frontal gyrus as the centre responsible for number pronunciation, the angular cerebral gyrus and the intraparietal sulcus as the areas of the brain responsible for number reading, and the angular gyrus for number writing). Hans Berger (1926) took up his ideas and further developed them, introducing the division between primary and secondary acalculia.

Primary or "pure" acalculia corresponded to the loss of numerical concepts and the inability to understand or perform arithmetical operations, while secondary acalculia referred to calculation imperfections due to other cognitive disorders (e.g. memory, language, etc.). Later on, the distinction became more important, because researchers generally agreed on the fact that calculation disorders were related to other cognitive disorders, such as aphasia, alexia, and dysgraphia. The dispute hence hinged on the possible existence of primary acalculia, because several authors questioned its existence as self-standing cognitive disorder (Collington et al. 1977; Goldstein 1948). In 1976, Alexander Luria (Luria 1976) pointed out the difference between optical acalculia (visual-perceptive), frontal acalculia, and primary acalculia, emphasizing the fact that calculation disorders could be caused by differed brain disorders. The author contended that calculation disorders were not homogeneous and, consequently, that it was reasonable to differentiate between different subgroups of acalculia. Several categorisation attempts were proposed (Grafman 1988; Luria 1973) and different error models were described in patients with damages to the right and left hemisphere (Rosselli and Ardila 1989). Similarly, Henry Hecaen and colleagues (Hecaen et al. 1961) classified different forms of acalculia on the basis of their main features, differentiating calculation from the processing of numbers and putting forward a new classification model, later adopted by Geary (Geary et al. 2000). On the basis of a study performed on 183 patients, Geary identified three main types of calculation disorders:

(1) dyslexia and dysgraphia for numbers;
(2) spatial acalculia;
(3) anarithmetria (primary acalculia).

Dyslexia and dysgraphia for numbers entailed patent calculation problems. These issues might or might not be related to the same disorders for words. Spatial acalculia referred to a spatial organisation disorder, which implied the inability to correctly apply the rules to place written numerals in the right order and position (often leading to numerical inversions). Anarithmetria (also known as primary acalculia) entailed fundamental calculation disorders. Patients affected by anarithmetria are completely unable to understand quantities, struggle in the use of calculation syntactical rules, and have problems in the understanding of arithmetical symbols. Nevertheless, these patients can count out loud and perform certain tasks linked to numerical calculation (for example, the use of multiplication tables). They can also have numerical knowledge, despite being unable to compare numbers (quantity estimations) (Ardila and Rosselli 1995).

Conversely, it is quite difficult to find pure examples of anarithmetria in academic literature. Often enough, patients (as in previous cases) presented a general cognitive deterioration and, therefore, anarithmetria was easily related to other neurological and neuropsychological disorders.

Hecaen himself described cases of combination of anarithmetia and dysgraphia for numbers. In a sample of 73 patients affected by anarithmetia, the authors noticed that 62% suffered also from aphasia, 61% made construction mistakes, 54% had visual disorders, 50% had general cognitive disorders, 39% had alexia problems and 33% had oculomotor disorders.

There was a certain degree of overlap of these subtypes of acalculia and the author postulated that calculation skills represented a complex type of cognition that required the involvement of several cognitive skills.

Boller and Grafman (Boller and Grafman 1983) postulated that calculation skills could be damaged as a result of different types of disorder. Calculation skills could be affected by (1) the inability to understand the name of numbers, (2) visual-spatial disorders that inhibit the spatial organisation of numbers and the mechanical aspects of operations, (3) the inability to remember mathematical factors and use them properly, (4) the inability to have mathematical thoughts and follow basic functioning rules.

Brain damage can therefore entail limited disorders related to specific arithmetical functions without causing additional calculation disorders (e.g. a specific disorder in arithmetical procedures). The first studies inquiring the similarities between the characteristics of mathematical disorders in childhood and symptoms of adults with acalculia (or acquired dyscalculia) were carried out at the beginning of the 19th century. According to the DSM-IV (*American Psychiatric Association*), Specific Calculation Disorder (or Dyscalculia) is diagnosed when children obtain lower marks in standardised mathematical tests compared to the expected results for their age, intellectual level, and education.

Specific learning disorders in mathematical operations are not so uncommon among children. These disorders are recorded also without clear signs of linguistic difficulties or with normal IQs. It has been estimated that almost 6% of school children have some kind of difficulties with mathematics, with a 70% incidence among males (Badian 1983). Nevertheless, the research has not yet been able to provide a common method to analyse the typologies and causes of the difficulties related to calculation disorders. As in the case of acalculia, several categorisation models have been presented also for dyscalculia. The earliest example dates back to 1967 and was developed by Johnson and Myklebust (Johnson and Myklebust 1967). These authors analytically identified different types of calculation difficulties in their patients, such as establishing a one-by-one correspondence between elements, being unable to recognise the relationship between a symbol and a quantity, being unable to associate auditory symbols (name of numbers) with visual symbols; being unable to learn the cardinal and ordinal system of enumeration and counting; being unable to understand the principle of quantity conservation, being unable to perform arithmetical operations, and finally being unable to understand the meaning of the symbols used in operations. According to the authors, mathematical disorders rarely appear in their purest form, while they are normally associated with other specific developmental disorders, such as reading, motor, or attention disorders.

Starting from this detailed list of difficulties in calculation disorders, manifold definitions of developmental dyscalculia have been formulated in academic literature, referring to specific hypotheses of cognitive disabilities. Robert Cohn (1971), for example, defined dyscalculia as a retardation in numerical skills development, characterised by the inability to recognise numbers, remember basic operations (multiplication tables) and keep numbers in the right order during calculations. Nathlie Badian (1983) took the categories put forward by Hecaen (alexia and/or agraphia for numbers; spatial acalculia, anarithmetia) as a starting point and added a fourth category: attentional-sequential dyscalculia, which entails difficulties in performing additions and subtractions, remembering multiplication tables, and remembering carry-overs and commas. Also Christine Temple (1991) distinguished three types of dyscalculia:

(1) Numerical dyslexia, a disorder characterised by difficulties in acquiring lexical procedures both in the systems needed to understand numbers and make calculations.
(2) Procedural dyscalculia, which—contrary to numeral dyslexia—is typical of children who struggle to learn the procedures needed for calculations, despite the fact that they do not have problems in reading and writing numbers.
(3) Dyscalculia for arithmetic facts. In this case, despite owning a good ability in processing numbers and perform calculations, children struggle to learn the numerical facts within the calculation system, making so called "border mistakes" (mixing up multiplication tables, e.g. $6 \times 3 = 21$) or "skidding mistakes" (writing one wrong digit in the answer, e.g. $4 \times 3 = 11$).

Temple's studies described the possible types of developmental dyscalculia by characterising them according to the causes and neuropsychological preconditions of the disorder. Nevertheless, it is worth noting that not even today is there a shared modality among researchers to analyse the causes of the difficulties implied by calculation disorders. Generally speaking, dyscalculia is always associated to dyslexia (Temple reported the case of a patient suffering from phonological dyslexia and dyscalculia in multiplications). Nevertheless, in a study performed on children affected by dyscalculia, Geary, Hamson and Hoord (Geary et al. 2000) highlighted that there is a difference among subjects affected by dyslexic dyscalculia and those affected by a form of dyscalculia not associated to dyslexia. In their experiment, when faced with additions, children with dyslexia often counted on their fingers to find the right result, while children without dyslexia never used this strategy and always tried to recall in their memory the addition tables they needed. This fact was explained by putting forward the idea that when children suffer from a form of dyslexia associated with dyscalculia, they do not have the possibility to represent numbers verbally and therefore avoid searching in their memory a result that is simply impossible to remember for them. On the contrary, children without dyslexia do not have memorisation problems related to multiplying tables. It is therefore likely that their disorder is due to a representation problem related to quantities. In this case, the strategy of counting on fingers makes no sense.

Geary's research supported the hypothesis of a hierarchy in strategy learning. Initially, children use a procedure called *counting all*, counting addends on their fingers by raising them one after the other (to add up 4 and 3 they lift four fingers, then three, and then they count the total). At the end of the first school year, they start using a strategy called *counting on*: they start counting from the highest addend and add the smallest one, one unit at a time. The ultimate strategy requires them to look at their fingers without counting and find the answer. Counting on fingers in arithmetic therefore represents an important development stage of numerical cognition, so much so that Fayol, Borrouillet and Marinthe (Fayol et al. 1998) have proved, in an experiment on a sample of 200 normal children, that the level of ability of 5-year-old children in finger counting is a far better index of their arithmetical performance at 8 years than their IQs.

The high number of theories and models outlined so far suggest that even the definition of dyscalculia is not always univocal. What is puzzling is that often the neuropsychological models presented to study single cases of known cerebral damages are also used to define the characteristics of dyscalculia as a specific learning disorder developed by children while growing up. These classifications have provided a wide framework of possible manifestations of this disorder, but they merely play a descriptive role.

Only after formulating cognitive models that describe the processing of numbers and calculation, the functional meaning of each observed disorder *pattern* becomes fully understandable.

4.1.1 Calculation Disorders and Cognitive Models

Michael McCloskey, Alfonso Caramazza and Annamaria Basili (McCloskey et al. 1985) developed one of the earliest cognitive models of numerical processing and calculation. In it, the authors generally considered the cognitive mechanisms that lead to the understanding and production of Arabic and verbal numerals, as well as the execution of simple calculations. According to their model, the mental representation of numerical knowledge is independent from other cognitive systems and structured in three modules with different functions: (1) the system for number understanding and production (divided in the different codes in which numerals are encoded: verbal, Arabic, etc.), (2) the calculation system (divided, once again, in recovery systems for arithmetical facts and calculation procedures), and (3) the production mechanisms, the *output* of the calculation system, i.e. numerical answers.

This implies that the understanding of numbers and their production are two processes independent of each other. The units tasked with the processing of numbers in different formats are independent as well, both in the understanding and the production of numbers. The recognising/understanding system is composed of a series of independent mechanisms, for example those used to process numbers as the Arabic numeral "8", the visual and verbal number "EIGHT" and the auditory and verbal number "eight". Also the production system has a complex internal structure that comprises mechanisms responsible for the production of the Arabic numeral "8", the oral production of "eight" and the written word-number "EIGHT", mechanisms that work independently of each other. Let's see an example of the use of this model to represent the solution of a simple arithmetical problem: "8 × 3" or "eight times three".

(1) The problem is presented:

   (a) in Arabic numerals and mathematical symbols: 8 × 3;
   (b) in written or oral form: "eight times three".

(2) The problem is correctly understood by using the appropriate understanding mechanism (decoding of numerals or words) and is converted into an abstract semantic code.
(3) The abstract representation is used to access the calculation mechanism: arithmetical facts or procedures/algorithms.
(4) The calculation mechanism employed provides an abstract semantic representation.
(5) The representation is sent back to one of the two production mechanisms and is expressed in:

   (a) Arabic numerals and mathematical symbols: 24;
   (b) written or spoken words: twenty-four.

Among the lexical components for understanding and producing verbal numbers, the model differentiates between phonological processing mechanisms for spoken words and graphemic processing mechanisms for written words. The model does not postulate a phonological-graphemic difference for the processing of syntax and it is believed that the same syntactical mechanisms are used for processing spoken and written verbal numbers (Fig. 4.1).

A salient feature of this model is that it foresees possible alterations and disorders in the processing of numerical information. This model has been entirely developed to reunite different neuropsychological studies that highlighted alterations of each aspect linked to numerical and calculation processing (McCloskey et al. 1991). Several studies have pointed to the decoupling of understanding and production mechanisms. Benson and Denckla (1969), for example, reported the case of a man with a damage to the left hemisphere: the subject was able to choose

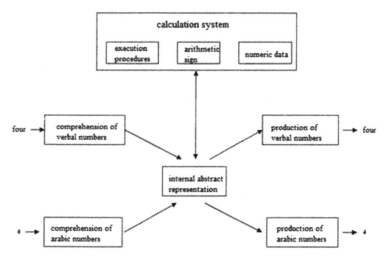

Fig. 4.1 The modular model developed by McCloskey et al. (1985)

the right answer in a list of possible options when a number was spoken, hence showing that he could understand numbers in different presentation modalities (visual and auditory).

Nevertheless, when the patient had to produce a number, his performance deteriorated significantly until failing to solve even simple arithmetical tasks. At the same time, he made obvious mistakes both in writing and reading numbers. Ferro and Botelho (1980) instead described the case of two patients (AL and MA), who presented selective disorders in the understanding of arithmetical symbols. When they were presented with written arithmetical problems, both patients made mistakes in the operation. For example, AL answered 12 when he was presented with the addition 3 + 4. The interesting fact is that both patients were successful in processing numbers and gave the right result when the operation was presented orally to them, since they did not have problems in understanding the words "plus" or "times".

Similarly, the patient VO of Michael McCloskey and colleagues (McCloskey et al. 1985) showed full understanding of numbers and calculations presented verbally and in the form of Arabic numerals, succeeding in indicating without difficulties the largest number between two alternatives and the number representing the right answer to an operation, but on the contrary in the tasks that required number production skills to answer a stimulus, his performances were strongly compromised. The patient made blatant mistakes in solving mental and written calculations, writing spoken numbers, and reading numbers out loud, showing a decoupling between the ability to understand numbers (intact) and produce numbers (damaged). As far as comprehension is concerned, the authors reported the case of patients presenting a double decoupling of verbal and numerical codes. In a test requiring the ability to identify the largest number between two alternatives presented visually, the patient H.Y. did not struggle in identifying the largest number when Arabic numerals were presented (e.g. 8–5; 27305–27350), but he answered randomly when numbers were presented as words (EIGHT-FIVE, etc.). Conversely, patient K. presented the opposite behaviour: he did not have problems with numbers presented as words and provided random answers when numbers were presented as numerals.

This double decoupling has also been confirmed by other studies. For example, patient S.F. of Lisa Cipollotti (Cipollotti et al. 1995) could read correctly 95% of numbers (from 1 to 7 numerals) written as numerical words, but only 45% of the same numbers written as Arabic numerals. Similar problems, despite a much more severe disorder than in the previous case, were presented by patient BAL (Cipollotti et al. 1995). BAL could read correctly the words corresponding to the numbers from one to nine, but gave wrong answers when the same numbers were presented as Arabic numerals.

On their part, Clark and Campbell (Clark and Campbell 1991) put forward a different hypothesis to the modular model. They believed that the different association, semantic and linguistic mechanisms involved in calculation processes are so deeply interconnected that the modular model seems unlikely. This theory, also known as "theory of complex encoding", postulates an equipotential explanation between the different functions. In other words, according to the authors, calculations cannot be broken down into functional and/or anatomical categories. This model outlined a non-modular architecture where multiple numerical codes are triggered one after the other during the processing of numbers and arithmetical tasks. According to this hypothesis, the numerical codes include phonological, graphemic, visual, semantic, lexical, articular, imaginative, and analogical representations. Numbers automatically activate a wide network of associations and, in the context of a specific task, include both relevant and irrelevant information. For example, when subjects are asked to quickly solve simple additions or multiplications, the mistakes they make reveal the excess of influences: mistakes are normally due to the recovery of "similar" numbers from an associative or semantic point of view (e.g. $3 \times 6 = 21$), or for computing a wrong arithmetical operation associated to the problem (e.g. $3 \times 6 = 9$).

The theory of complex encoding tries to provide a simplified theory compared to the modular model. This theory has the merit of having simple theoretical foundations that explain several phenomena observed in clinical and laboratory works. Unfortunately, this is not true on a deeper level: this model is too generic and it has the disadvantage, in Popper's terms, of not being falsifiable. The need to go back to a modular conception adapted, if adapted to the hypotheses of complex encoding, was heeded by Stanislas Dehaene, who developed a triple-code model that identified three categories of mental representations where numbers could be manipulated (Dehaene 1992; Dehaene and Cohen 1995). This model aims at better defining the numerical part of the processing tasks, postulating that three cardinal representations are sufficient. The first mental representation category is the visual format (Arabic) of numerals. At this representation level, there is an ordered list of numerical entities (e.g. 12 is encoded as <1> <2>). The second category is given by the verbal structure in which numbers are ordered as sequences of syntactically organised words. At this representation level, the number "twelve" is encoded as "Ten[1] Units [2]". In these first two categories, no semantic information is stored, while it is provided by the third and last category of the model, the analogic representation. At this level, a connection is made between the quantity or associated size of a specific number and other numerical quantities. Neuropsychological experiments were carried out to prove the efficiency of this model. The most famous case is described by Dehaene and Cohen (Dehaene and Cohen 1991) and refers to patient NAU, who had a diffused lesion to the left hemisphere and presented moderate disorders in the spoken language. Furthermore, he could not read nor write. In a test on word categories, the only words he could read and write were numbers. Furthermore, despite not being able to write letters when they were spoken to him, he was able to write 7, 43, 198, and 1985 to answer the stimuli 7, 42, 193 and 1865 (it must be noticed that written numbers are not so distant from the spoken stimuli). In particular, what seemed relevant in the performance of patient NAU is that he succeeded only in approximation tasks: he could easily identify false results in additions (e.g. 2 + 2 = 9), but he even accepted results that differed from the real result for just few units (2 + 2 = 5) (Fig. 4.2).

When asked to repeat a sequence of numerals, such as (6, 7, 9, 8), the patient was only able enumerate the list for few minutes. In addition to this, he could not tell if a number was near to the ones presented in the list (for example, 5), but if he was presented with a distant number (e.g. 2), he could tell that the number did not belong to the list. The performances of patient NAU were therefore sensitive to the numerical distance between stimuli: stimuli that were too near became undistinguishable. For NAU, numbers were not accurate as they are for normal people and each numeral recalled just a vague impression of quantity.

A dyslexic patient described by Laurent Cohen and colleagues (Cohen et al. 1994) presented a similar behaviour. In this case, despite the patient strongly

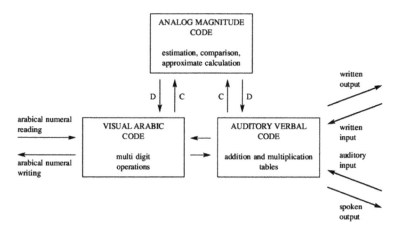

Fig. 4.2 Dehaene's triple-code model

struggling with reading numbers out loud, he was able to count numbers with one or two digits. Furthermore, when complex yet familiar numbers were presented to him, such as famous historical dates or postal codes, the patient could understand their right meaning.

In conclusion, we can state that providing the proper solution to a numerical problem requires verbal, spatial, and conceptual skills that, very likely, require the active involvement of several cerebral structures. Nevertheless, the neuronal mechanisms involved in number recognition seem to be different to those involved in solving arithmetical problems. A patient may indeed struggle with number recognition and, simultaneously, easily solve arithmetical operations. Therefore, in the following paragraphs, we will investigate how brain-imaging techniques can help identify the areas in the brain devoted to the processing of arithmetical tasks.

4.2 Experimental Evidence

Considering the substantial progress made over the last years by neuroscience, it was obvious that neurobiologists and neuroscientists would focus their attention on mathematics, particularly on the production and understanding of numbers seen as activities deeply anchored in the brain. Despite the lack of an unequivocal and unambiguous interpretation of experimental data and evidence, there is no doubt that the efforts made in neuroscientific research represent a new and innovative chapter in the study of numerical cognition. Neuroscientists always build their conclusions on what they consider a *corpus* of convincing evidence obtained through a series of reliable experiments.

The majority of these tests are based on functional imaging studies analysing brain activity during mathematical tasks. The use of non-invasive techniques for the monitoring of brain activity such as PET (*Positron Emission Tomography*) and fMRI (*Functional Magnetic Resonance Imaging*) has indeed allowed for the observation of changes in brain activity when performing different activities, like quantifying stimuli, performing simple operations, recognising numerals and translating them in internal quantities, or detecting the location and extension of brain damages related to calculation skills.

In 1985, Roland and Friberg, two forerunners in the use of imagery in the study on calculation, used PET to record the activation of a wide frontal-parietal region while the subject mentally performed a sequence of subtractions. Roland and Friberg, whose ambitions went beyond the localisation of calculation skills, contended that they had found the basis of human symbolic reasoning in this region (Roland and Friberg 1985). Ten years later, when brain-imaging techniques had developed even more, becoming easier to use and less invasive, the cerebral region where PET studies (Ghatan et al. 1998; Jong et al. 1996) had recorded activation was also analysed with a Functional Magnetic Resonance (Burbaud et al. 1995; Burbaud et al. 1999; Rueckert et al. 1996), confirming the earliest results. Nevertheless, these authors also used a series of complex cognitive tasks that, in addition to numerical encoding, required the involvement of other cognitive mechanisms to identify the stimuli in their studies. These early studies did not have the merit of specifying the role played by each cerebral region in the identified functional network, but they provided inspiration for a second generation of researchers that tried to define a controlled task that could single out the cognitive process under investigation. Simultaneously, the development of image acquisition and statistical analysis techniques promoted the establishment of richer experimental protocols that could compare different experimental settings. The earliest studies, in line with the neuropsychological observations made by Gerstman, Hecaen and Henschen, emphasized the involvement of the parietal lobe in numerical tasks. Recent results instead pointed to the horizontal part of the HIPS parietal lobe (i.e. "horizontal part of the intraparietal sulcus" located at −44 −48 47 right, 41 −47 48 left).

At the beginning of 2000, a series of studies (where cerebral activations linked to different arithmetical operations were measured and compared) confirmed that *HIPS* is more active when calculations are performed rather than when numerals (Burdaud et al. 1999; Chocon et al. 1999; Pesenti et al. 2000) or letters (Simon et al. 2002) are simply read, while it is even more active when two calculations are involved (Menon et al. 2000). Furthermore, among different arithmetical operations, it seems that the activation is stronger in the case of subtraction rather than in the case of simple comparisons (Chocon et al. 1999) or multiplications (Chocon et al. 1999; Lee 2000). As far as additions are concerned, the *HIPS* seems more active in approximation tasks rather than exact ones (Dehaene et al. 1999). Similarly, Manuela Piazza and colleagues (Piazza et al. 2002) proved that *HIPS* is activated not only by symbolic tasks, but also by those where it is necessary to estimate the numerosity of groups of dots. All these data confirm the higher level of activation of *HIPS* when a task involves the manipulation of quantities (subtraction, addition, numerosity approximation and estimation) and numerical comparisons (Chocon et al. 1999; Pesenti et al. 2000).

Comparison tasks provide instead the perfect opportunity to observe in detail the parameters affecting activation. In an fMRI study, Philippe Pinel (Pinel et al. 2001) observed that the activation of *HIPS* was affected by the distance effect: in other words, *HIPS* was more active when the numbers under comparison were nearer.

The author aimed at proving that only numerical distance had an effect on the activation of *HIPS*. In particular, the activation level was the same when stimuli were presented as Arabic numerals or in the form of written words. In a 1998 study, Naccache and Dehaene (Dehaene et al. 1998; Naccache and Dehaene 2001; Dehaene 2011; Dehaene et al. 2003) proved that reading Arabic numerals immediately recalls the associated quantity, even when numerals are presented only briefly and subtly. The subjects tested in their study had to classify numbers ("the objectives", all clearly visible) establishing if they were larger or smaller than 5. The subjects were not aware of the fact that before each objective they were subliminally presented with other numerical stimuli ("the baits"). The answer times of subjects were faster when the two numbers were associated to the same answer (both numbers larger or smaller than 5) and this was recorded with all the modalities used as objective or bait (numerals or words). At the same time, the data obtained through fMRI highlighted a lower *HIPS* activation when the bait and the objective were numerically identical, while the activation was stronger when they were different.

Obviously, since subjects in these experiments were exposed to different paradigms that entailed various tasks, non-numerical counterarguments can be presented: it is always possible to try finding an explanation to results by referring to general mechanisms related to attention or the planning and execution of tasks. Evelyn Eger and colleagues (Eger et al. 2003) tried to avoid this trap by measuring the cerebral activity during the execution of a task requiring to identify the objective (for example, identify the number 2 in a sequence of numbers presented one after the other on the screen), noting that subjects solved this task both with numbers and letters (visually presented on the screen or read out loud by an auditory stimulus) and colours (a coloured rectangle on the screen or the name of a colour presented in auditory form). The three tasks had the same difficulty level and involved both the attention and execution skills of subjects. By comparing the *HIPS* activation levels for numbers with those for letters and colours, the authors systematically and exclusively identified two symmetrical regions of the *HIPS* located in the same areas as those highlighted by previous studies. These results hence prove that the activation curve of this cerebral area is independent of the presentation modality of stimuli and identifies the cerebral areas that abstractly encode numbers in the HIPS of the two hemispheres. As we have seen, there are studies that describe a bilateral activation in the intraparietal sulcus. How do these bilateral activations work? Do these regions play an indispensable role in encoding numerical quantities? Do they work in combination or are they independent of each other?

Experimental data on patients suffering from a detachment of the corpus callosum show that the two hemispheres can solve comparison tasks independently (Gazzaniga and Hillyard 1971; Gazzaniga and Smylie 1984; Seymor et al. 1994; Cohen and Dehaene 1996). As a matter of fact, when a numeral is presented for a brief period of time in the field of vision of each hemisphere in patients without the corpus callosum, the perception and processing of stimuli are exclusively performed by the contralateral hemisphere. When numerals were presented to the left hemisphere, the patients behaved normally: the left hemisphere could solve numerical tasks autonomously. On the contrary, when numerals were presented to the right hemisphere, patients struggled to read them and their productions got near to the exact results without being entirely correct: the patients nevertheless kept the ability of making comparisons between numbers.

In this case, patients behaved as "approximate patients". To explain these data it was postulated that the right hemisphere is indeed responsible for the encoding of quantities, but rather approximately (Piazza et al. 2004). In conclusion, it can be deduced that both hemispheres can make representations of the quantities associated to numbers, even if the left hemisphere is better at managing exact numbers than the right hemisphere.

In the light of all this, what is still up for discussion is the role played by the right hemisphere in the selection of the exact answer. Some neuropsychological studies do not support the idea that the right hemisphere is responsible for this task. The previous hypotheses contended that manipulation skills on quantities can resist cerebral damages, because their alteration requires lesions affecting bilaterally both intraparietal sulci.

It is indeed unusual to find cases of patients suffering from disorders in the processing of quantities due to unilateral damages in academic literature. Up to now, only four cases have been recorded. The first patient with substantial disorders in tasks requiring the ability to process numbers bigger than 4 (Cipollotti et al. 1991) even with non-symbolic stimuli, presented a unilateral lesion in the left frontal parietal area. The second patient presented an unilateral lesion in the right lower parietal region (Dehaene and Cohen 1997). The third patient (Delazer and Bencke 1997) presented an unilateral lesion in the left parietal cortex. Finally, the fourth patient (Lemer et al. 2003) suffered a lesion limited to the left part of the parietal lobe in the intraparietal area. By examining these patients, it could be concluded that only the dominant hemisphere is necessary to solve quantitative tasks on numbers. Nevertheless, this observation must be verified through the study of patients with unilateral lesions (sometimes these lesions involve a large part of the left hemisphere) who are unable to perform tasks that represented a problem also for the above-mentioned patients (Deahene and Cohen 1991; van Harskamp et al. 2002). This fact could be explained by postulating that the patients of the first group were more strongly affected by lesions and even apparently healthy regions had lost some of their functionality.

This phenomenon can be caused by the decoupling of healthy and injured regions or by metastases (metastasis effect: a remote influence by a lesion; examples are hypometabolism in an apparently healthy region). In addition to the case of patient NAU described above, in academic literature there are cases of other patients that lost the ability to perform exact calculations after a more or less extended lesion to the left hemisphere (Dehaene and Cohen 1997; Grafman et al. 1989; Cohen et al. 1994; Warrington 1982; Pesenti et al. 1994). Besides, the same performance level was also recorded after a lesion to the corpus callosum, particularly when stimuli were presented to the right hemisphere (Gazzaniga and Hillard 1971; Gazzaniga and Smylie 1984; Cohen and Dehaene 1996; Seymour et al. 1994). In these patients, the two hemispheres worked independently and the performances of the right hemisphere were similar to those of the patient that had lost all functionalities of the left hemisphere.

In a study performed in 2003 by Cathy Lemer and colleagues (Lemer et al. 2003), the researchers developed a list of tasks in which approximation patients presented characteristic disorders. For example, after a frontal and temporal lesion of the left hemisphere, their patient BRI struggled with arithmetic. The tests revealed that he struggled the most with multiplications and divisions, while he did not have particular difficulties with additions and subtractions.

The few mistakes made by BRI in additions and subtractions concerned bigger numbers and his answers were always near to the right result. After these tests related to arithmetical operations, the authors performed other tests on BRI: firstly, a test of exact and approximate addition, where he struggled the most with exact problems with big numbers; then an estimation test on groups of dots, where the patient's performances did not reveal particular issues; finally, a comparison and addition test concerning groups of dots, where the patient's performances were quite similar to those of a control group (with the exception of a general delay due to his executive commands). BRI was therefore able to solve concrete and approximate arithmetical problems. He could also perform tasks of addition and subtraction with symbolic stimuli, but only approximately. On the other hand, BRI was not able to manage quantities in multiplications and divisions. In the cases described above, a "distance effect" was visible also in tasks where usually, in normal subjects, this effect did not appear.

This was due to the fact that these patients systematically use an analogical code of quantity to solve numerical tasks, a system that shows limitations in its accuracy (in the case of mister NAU, there were no problems in distinguishing distant numbers such as 4 and 9, while 4 and 5 were mixed; on the other hand, BRI did not have particular problems in comparison tasks, even in the case of small numbers). The answer accuracy varies according to the patient. There are even patients that have completely lost the sense of quantity. This is the case of LEC, a patient described in the same paper reporting the case of BRI. After a lesion to the left intraparietal area, LEC struggled with arithmetic. In a classic test of elementary operations, LEC showed disorders in the case of subtractions and divisions. The mistakes made (e.g. $7-1 = 8$; $9-1 = 9$) pointed to a limited understanding of the concept of subtraction. His performances were good in exact and approximate addition tasks, but he stated that he could solve these tasks following the same procedures, i.e. by using addition tables that allowed him to calculate the exact result. On the contrary, the tasks related to groups of dots emphasized the extension of his disorders. When asked to compare two groups of points, LEC answered randomly.

Generally speaking, comparison tasks implied more difficulties for LEC, who had problems also when comparing numbers between 20 and 100 written as Arabic numerals.

Other patients presented the same selective disorder in subtractions as LEC (Dagenbach and McCloskey 1992). An example is JG, a patient analysed by Delazer and Benke (Delazer and Benke 1997), who was able to perform multiplications (8% of error rate on elementary facts), but was at the same time unable to understand the meaning of numbers and operations. Also Dehaene and Cohen (Dehaene and Cohen 1997) presented the case of a patient (MAR), who was able to do multiplications but could not compare or estimate the result of an operation or find the middle point of a numerical segment. Finally, Margarete Delazer and colleagues have reported the case of a patient with similar disorders to those presented by LEC (Delazer et al. 2005): this patient could perform multiplications and additions, but struggled with subtractions. In general, his performances were very low in all tasks requiring a systematic processing of quantities (bisection of numerical intervals, approximation, estimation, placing numbers of a physical line).

4.3 Calculations and Genetics

In the previous pages, we outlined experimental evidence of the numerical quantity, non-verbal representation system. Furthermore, it was highlighted that this system—which develops quite early—can be seen as the foundation on which arithmetical competences are built. Over the last years, this research activity triggered a new series of studies focussed on the investigation of the development of these competences in children with atypical development patterns. The discovery in 2001 that the mutation of a single gene (FOXP2) could have an impact on language learning paved the way to this kind of studies (Laing et al. 2001). The localisation of FOXP2 was linked to the discovery of a particular linguistic disordered identified in one family (called KE): half of its members presented a grammar deficit that was particularly visible in the use of linguistic features signalling number, gender, and time. All the members affected by this disorder presented the mutation of gene FOXP2 in chromosome 7, while the members of the family immune to linguistic disorders did not have this mutation. Nevertheless, the consequences due to the mutation of FOXP2 were particularly complex to study, because these genes are involved in several functions and their role is not exclusively linked to language learning. Today, we know for sure that FOXP2 governs the activities of other genes and, consequently, affects the development of several organs. For example, it controls the production of a protein (*forkhead protein*) that plays a fundamental role in the development of proper motor and coordination abilities, as well as in the production of verbal and non-verbal rhythmic sequences. Nevertheless, it is believed that a better understanding of its role might lead to an improvement in the comprehension of the variability of cognitive functions in normal individuals as well as linguistic disorders in subjects with disorders.

The same goes for the study related to the exact and approximate numerical system in atypical development patterns. Annette Karmiloff-Smith (1998) rightly highlighted that the studies exploring the field of numerical cognition should not solely take into account the final result of development and all possible disorders, but rather focus on the inquiry and understanding of the development patterns of the different cognitive processes involved. From this point of view, by focusing on a homogeneous neurobiological phenotype, the study of genetic conditions can provide the possibility to evaluate arithmetical competences in the perspective suggested by Karmiloff-Smith. Furthermore, issues of dyscalculia have often been reported in association with genetic disorders such as Williams syndrome, Down syndrome, fragile X or Turner syndrome.

Williams syndrome (WS) is a very rare genetic disorder that occurs in one on 20,000 births. The genetic causes of this disorder are due to the micro deletion of the elastin gene in the chromosome 7. The clinical picture of the subjects affected by WS is characterised by a dysfunction of several organs and systems: they suffer from cardiovascular, renal and auditory problems. Furthermore, they present visual-motor and visual-constructive difficulties (they are not able to manipulate single spatial pieces of information and place them in a coherent context), while their linguistic abilities seem unaffected. When talking to subjects with WS, the first impression is that they are eager talkers, but after a deeper exam it becomes clear that it is not true. Their presumed linguistic ability is simply due to a good phonological memory that helps them to learn sounds, words, or sentences that are mainly copied. These subjects therefore often know many words, even uncommon and sophisticated ones, but they do not master them with full semantic competence and often their disorders are of pragmatic nature. In several studies, subjects affected by WS have been compared with individuals affected by Down syndrome. Sarah Paterson and colleagues (Paterson et al. 1999; Paterson 2001) did so in their studies on numerical cognition, proving that Williams children have excellent approximation skills and can perceive quantities already at an early age. On the contrary, these skills are not recorded in children affected by Down syndrome. The most interesting aspect of these studies is that in older children and adults, the situation seems to be the opposite and better approximation performances are recorded in Down subjects compared to people affected by WS.

These results would corroborate once again the importance of studying numerical competences in a dynamic and non-static approach to the disorder. The problem of numerosity comparison was also addressed by performing experiments on children and adults with Down and Williams syndrome, putting them in comparison with a control group of mental age peers (Paterson et al. 2006). The subjects with WS behaved with less accuracy than those with Down syndrome and did not show a good distance effect in relation to their reaction times. The data pointed to some anomalies in the knowledge of the number line, a hypothesis corroborated by the observation of a consolidated ability of numerosity comparison already at the age of 5 in a control group of children with typical development, and a low level of similar ability in a group of children with WS of an average age of 7. Other authors showed interest in the ability of assessing the magnitude of different numerosities in this population and proved that the subjects struggled in making representations of magnitude: the magnitude judgment abilities of children with WS did not change from school to adult age, contrary to what happened in children with typical development, where a clear improvement was recorded from pre-scholar age, to scholar age and adult age (Ansari 2003). There are also interesting studies that evaluated the skills on exact numerical systems, in particular counting skills and the learning of the principle of cardinality.

The data show that children with WS could enumerate sequences of small numbers while counting without making mistakes, while on the other hand they struggled with the principle of cardinality (Ansari et al. 2003). When asked "how many...?" they answered correctly less than 50% of times, showing that counting skills cannot predict the acquisition of the principle of cardinality in children with atypical development. Furthermore, they presented substantial difficulties in adult age as well: in particular, while the enumeration from 1 to 20 was learnt without substantial problems, several difficulties were recorded when counting from 25 to 35. These data pointed to the fact that these people learn counting as a form of nursery rhyme that does not require the acquisition of the concept of numerical line (Paterson et al. 2006). Children and adults with WS presented also reading diffi-culties for numbers with several digits (in particular, they made many lexical and syntactical mistakes), contrary to adults with Down syndrome, who showed to have a good competence of the transcoding process (Paterson et al. 2006).

The mentioned studies on Williams syndrome are quite interesting because they show that numerical competences can change during the development of the sub-jects. These data show that the relations between language and number are not defined once and for all, but they are in continuous evolution during development. A different pattern is visible in children with Down syndrome.

In their early years, these children show several difficulties in the perception of changes in numerosity compared to their peers, while older children and adults have a better ability in exact numerical tasks compared to children with WS of the same mental age (Paterson et al. 2006). This is an important aspect because it goes against what could be expected: if numerical skills were strictly linked to language and language were preserved in WS patients, we would expect better competences in these subjects rather than in those with Down syndrome. This observation is due to the fact that numerical competences related to exact systems do not depend solely on linguistic skills, but also on other abilities, *in primis* visual and spatial ones.

Furthermore, promising studies were performed on patients with Turner syndrome (Turner 1938), a genetic disease due to the presence of only one X chromosome. The disease affects one girl on 2,000, for a total of almost 3% of all female new-borns. The physical phenotype of Turner syndrome is well known and implies a small size, ovarian failure, and an abnormal development of the pubis accompanied by internal anomalies (e.g. hearth malformations). Contrary to the physical phenotype, the effects of the lack of one X chromosome on brain devel-opment are less known. Neuropsychological studies show that the cognitive profile of patients with Turner syndrome is characterised by a decoupling between insufficient non-verbal skills paired with normal verbal skills that often are even well above the average (Temple and Carney 1996). Even if the cognitive profile varies from one patient to the other, some disorders (visual- and spatial deficits, calculation and memory deficits) seem to be recurrent in subjects affected by Turner syndrome. The association of a visual- and spatial- disorder with dyscalculia is the potential effect of a close relationship between visual- and spatial- cerebral repre-sentations and numerical representations (Fisher et al. 2003), pointing towards a "spatial" dyscalculia related to an abnormal development of numerical represen-tation. Among patients affected by Turner syndrome, 75% struggle with mathe-matics, in particular with subtraction, operations with big numbers, and subitization. These difficulties related to mathematics can be found at all ages and across all social statuses (Temple and Marriot 1998).

In a study carried out by Nicolas Molko and colleagues (Molko et al. 2003) to investigate Turner syndrome, in the majority of cases patients presented calculation disorders, despite their excellent verbal IQ. By comparing the data of 14 subjects with Turner syndrome (TS) with those of 14 normal subjects (control group), the author reached the conclusion that the subjects with TS had significantly worse results than the control group in Warrington's arithmetical test. The average correct answer rate was of 41.1% (vs. 60.1% in the control group). In general, in numerical tasks related to large numbers, subjects with TS were slower and made more mistakes than the control group. A similar pattern was observed in the exact estimation of numbers (with an error rate up to 20.5%). On the contrary, the control group did not show a substantial difference in the test for exact and approximate calculation, neither in reaction times nor in error rates. Nevertheless, the format had a significant effect: when the number grew bigger, the answers slowed down by 231 ms, reaching an error rate higher by up to 9% compared to smaller numbers.

The data collected by Molko and colleagues through the use of fMRI to study the functional activation and the morphology of the intraparietal sulcus deserve a particular mention. The morphological analysis showed an abnormal length and depth of the right intraparietal sulcus that reflected an anatomical disorder of this cerebral region in subjects affected by TS. The data collected with the use of fMRI emphasized an abnormal modulation of the intraparietal activation according to the format of the number: in exact calculation tests, when the format increased, normal subjects recorded a bilateral increase in the activation of intraparietal sulcus. On the contrary, in the subjects affected by TS, there was no change in the activation level of the same regions. The format effect in exact calculation is due to an increasing difficulty in recalling the corresponding arithmetical facts in the memory and, therefore, to an increase in the use of alternative strategies that involve quantity manipulation (Lefevre et al. 1996). The analysis of the activation in cerebral areas during exact and approximate calculations with small numbers also pointed to an abnormal model of parietal activation even for simple functions. In normal subjects, there was a stronger bilateral activation of the intraparietal sulcus for approximate tasks compared to exact tasks. This is in line with the hypothesis that small arithmetical calculations (2 + 3) are recalled from memory and do not require the manipulation of quantities (Stanescu-Cosson et al. 2000). In subjects affected by TS, on the contrary, there was no difference between exact and approximate calculations: these subjects did not solve these operations by fetching the results directly from their memory, but rather they used the same strategies employed for approximate calculations. Summing up, the evidence collected through the use of fMRI showed that, despite a partial compensation, arithmetical difficulties in subjects affected by TS do not only affect calculations with bigger numbers, but also basic arithmetical knowledge that involves numbers smaller than five (Fig. 4.3).

The insufficient use of the intraparietal sulcus in subjects affected by TS seems to be the direct cause of their arithmetical difficulties and it suggests a lack in their selection and/or execution processes. This fact has also been reported in subjects with fragile X syndrome (Rivera et al. 2002). These results therefore lead to the conclusion that the explanation of these problems has to be found in a biological disorder, since what is missing is the classic "elementary sense" of numbers.

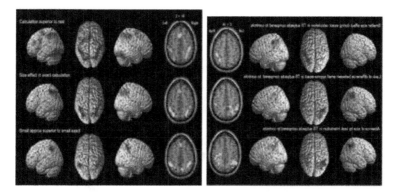

Fig. 4.3 Anatomy of the intraparietal sulcus in approximate calculations and exact calculations in normal subjects and subjects affected by TS (Molko et al. 2003)

4.4 Numbers and Space

In this book, the possible close relationship between space processing and number processing has often been highlighted. Even in the past, mathematical and spatial skills were often put in relationship: back in 1880, Galton was the first to openly refer to the "shape of numbers", hinting to the answers provided by 80 subjects to describe what they could "see" when they watched, heard, or conceived a number (Galton 1880).

The earliest encoding models of numerical quantities were direct products of the accumulator model developed to represent the duration of time (Gibbon 1977). The idea that there is a shared format to represent numbers and the duration of time represents an idea still cherished today, because it establishes a bond between numerosities and an internal magnitude (Walsh 2003). The most recent models on numerosities bring together all these ideas in the basic concept of "internal numerical line" (Dehaene 2011; Gallistel and Gelman 2000; Dehaene 2003).

According to this model, numbers (and numerosities) are represented on a mental line, with 1 standing on the left, 2 on its right, 3 on its right, and so on. To decide which number is larger between two alternatives, we start by placing them on our mental line and then we verify which one is standing more to the right. Numbers are not homogenously spread along this line. The more we move to the right in our mental line, the nearer the numbers stand (Zorzi et al. 2002) (Fig. 4.4).

Therefore, the metaphor of the mental numerical line described above is an adjusted format to describe the features of analogical, non-verbal representation skills: a practical tool to dwell upon these representations and understand their estimates. Interestingly enough, a certain number of evidence in neuropsychology suggest that the numerical line metaphor can go beyond its limits, providing important insights to understand the processing of numbers and space.

Some people develop explicit associations between numbers and space (Seron et al. 1992): for them, numbers stand on a line, on a curve, in a table, or follow

Fig. 4.4 The mental
numerical line (Galton 1880)

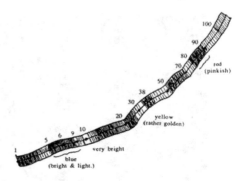

complex three-dimensional shapes (in a phenomenon known with the name of synaesthesia). Even "normal" subjects, though, can develop specific inferences between numbers and space. One of these effects is the so-called SNARC effect (*Spatial Numerical Association of Response Codes*), which unconsciously associates small numbers with the left part of the body and big numbers with the right (Dehaene et al. 1993).

A classic task to understand if we are affected by the SNARC effect requires to classify a number as even or odd by pushing a button either with our right or our left hand. In this test, numbers from 1 to 9 are presented one by one on a screen and subjects normally answer faster for numbers from 1 to 4 when they must press the button with their left hand, while from 6 to 9 they are faster with their right hand. The subjects hence respond as if numerals were categorised in smaller and bigger entities, despite the fact that the experimental task did not appeal to this notion of quantity.

In this way, the observations made on "normal" subjects are in line with those made on synesthetic subjects, indicating that it is possible to move one's attention on their mental numerical line. Some people even visualise their own body on the line. The numerical line of people affected by synaesthesia, just like the one of normal subjects, is an object that is external to the body and whose orientation and position are independent of the disposition of the body.

In the case of "normal" subjects, the orientation of this line is affected by culture: Iranians, for example, are used to writing from right to left and they will have a SNARC effect opposite to that of Western people. The interesting question is understanding whether spatial associations are an exclusive feature of numbers or whether they relate also to sequentially ordered non-numerical stimuli, such as the letters of the alphabet, the days of the week, musical notes, etc.

Wim Gevers proved in his study that the letters of the alphabet and the months of the year are also affected by the SNARC effect (Gevers et al. 2003), while musical notes can present the SMARC effect (*Spatial Musical Association of Response Codes*) (Rusconi et al. 2006) (Fig. 4.5).

The most striking demonstration of the equality in the processing of numbers and space, though, was provided by Marco Zorzi, Konstantinos Priftis and Carlo Umiltà (Zorzi et al. 2002). In their study, the researchers asked patients affected by

(a)

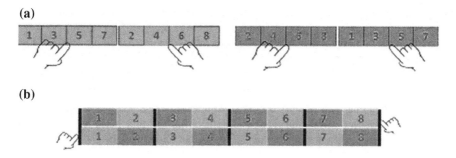

(b)

Fig. 4.5 Visualisation of the SNARC effect

neglect (an attention deficit disorder for the space located to the left and due to a trauma to the right parietal lobe that causes strange behaviours in patients, such as shaving only the right part of the face or eating only the right part of a portion of food, etc.) to indicate the middle point (bisectional numerical task) of a numerical segment. In general, the patients affected by neglect indicated a middle point that was placed much more to the right than the actual middle.

The same happened with numerical tasks: for an interval of numbers going from 11 to 19, the middle number for them was 17. The important fact in this study is that the same deviation recorded in bisectional numerical tasks was found in bisectional tasks of physical segments. If asked to point to the middle point of a line drawn on a piece of paper, these patients always point to a place standing too much to the right, making an error that is proportional to the total length of the line. These data prove that the same mechanism originally devoted to the orientation of attention in space is used to solve tasks of numerical bisection.

Laura Zamarian and colleagues (Zamarian et al. 2007) have found the same bisectional mistake of numerical intervals towards the right in patients affected by neglect in the case of two-digit numbers. Furthermore, it is interesting to note that some studies have proved a decoupling of the performances of patients with neglect in the case of bisectional numerical tasks and equality judgment. In these experiments, patients made mistakes moving too much to the right in numerical bisections and showed clear examples of the SNARC effect (Cappelletti and Cipolotti 2006; Priftis et al. 2006), suggesting that patients with unilateral left neglect can semantically process Arabic numerals.

It must be noted that a research has found a disorder qualitatively similar to the one described in patients with neglect in numerical bisectional tasks also in patients affected by schizophrenia (Cavézian et al. 2007), corroborating the idea that the spatial representation of numbers is a dynamic process. Furthermore, it was proved that the abilities of spatial exploration on the left side in patients with neglect can be improved by recurring to therapy treatments with prisms (Rossetti et al. 2004).

The positive effects of prisms in therapies aimed at reducing spatial mistakes has also been highlighted in a study on "normal" subjects (Loftus et al. 2008), who were asked to estimate if the numerical distance from a central number was bigger on the right or on the left of a trio of numbers (for example, 16, 36, and 55). This experiment indicated that subjects overestimated the length of numbers placed on the left of the central number, falling victim of the so-called pseudo neglect phenomenon (Jewell and McCourt 2000).

The numerical line is therefore not a simply useful tool to sum up the properties of a non-verbal representation of numbers, but it represents the explanation model for phenomena that treat equally the processing of numbers and space. As a matter of fact, the relation between the representations of number and space is evident also in neurologically normal subjects, as shown by an experiment by Martin Fischer (2001), who asked subjects to indicate the middle point of numerical sequences made up of a series of small numbers (e.g. 111, 222, etc.) and large numbers (e.g. 888, 999) in bisectional tasks. The results highlighted that the subjects moved the centre of the sequences of small numbers to the left and the centre of the sequences of large numbers to the right. The same results were later replicated by Calabria e Rossetti (Calabria and Rossetti 2005) and by Maria Dolores de Hevia, Girelli and Vallar (de Hevia et al. 2006).

Summing up, the "distance effect", the "magnitude effect", the "SNARC effect", and spatial performances in numerical bisectional tasks in patients with neglect and normal subjects have all proved the existence of a mental numerical line as a preferred representation form used by humans to process and elaborate numerical information. What is still unclear is how these numerical representations are wired at a cerebral level. On the one side, there are the neurons dealing with numerical quantities, while on the other, those dealing with space. Considering this, it is likely that the neuronal cell codes used by these neurons are equivalent and that there are mechanisms designed to translate one code into the other. Nevertheless, how do these two equivalent codes manifest themselves in the brain?

One possible answer is that the very same neurons are active in both types of representation, i.e. networks of neurons establish dynamic systems whose characteristics are determined by the wiring features of neurons (axons and dendrites length, myelination, synapse density, electrical properties of each cell). In the case of numbers and space, the two postulated codes are functionally equivalent and they would be triggered to represent the same neurons. Philippe Pinel and colleagues (Pinel et al. 2004) have reported a partial overlapping activation for two numerical (numerical comparison) and spatial (physical format comparison) tasks.

These data suggest that the neuronal cells responsible for the encoding of numbers and space partially correspond.

References

Ansari, D. (2003). *Atypical trajectories of number development: The case of Williams syndrome.* London: University College of London, Institute of Child Health.

Ansari, D., Donlan, C., Thomas, M. S. C., Ewing, S., Peen, T., & Karmiloff-Smith, A. (2003). What makes counting count? Verbal and visuo-spatial contributions to typical and atypical number development. *Journal of Experimental Child Psychology, 85,* 50–62.

Ardila, A., & Rosselli, M. (1995). Spatial acalculia. *International Journal of Neuroscience, 78*, 177–184.

Badian, N. A. (1983). Dyscalculia and nonverbal disorders of learning. In H. R. Miklebust (Ed.), *Progress in Learning Disabilities* (Vol. 5, pp. 129–146). New York: Grune and Stratton.

Benson, D. F., & Denckla, M. B. (1969). Verbal paraphasias as a source of calculations disturbances. *Archives of Neurology, 21*, 96–102.

Berger, H. (1926). Uber Rechenstorunger bei Herderkraunkunger des Grosshirns. *Arch. Psychiatr. Nervenkr., 78*, 236–263.

Boller, F., & Grafman, J. (1983). Acalculia: Historical development and current significance. *Brain and Cognition, 2*, 205–223.

Burbaud, P., Camus, O., Guehl, D., Bioulac, B., Caillé, J. M., & Allard, M. (1999). A functional magnetic resonance imaging study of mental subtraction in human subjects. *Neuroscience Letters, 273*, 195–199.

Burbaud, P., Degreze, P., Lafon, P., Franconi, J. M., Bouligand, B., Bioulac, B., et al. (1995). Lateralization of prefrontal activation during internal mental calculation: A functional magnetic resonance imaging study. *Journal of Neurophysiology, 74*(5), 2194–2200.

Calabria, M., & Rossetti, Y. (2005). Interference between number processing and line bisection: A methodology. *Neuropsychologia, 43*, 779–783.

Cappelletti, M., & Cipolotti, L. (2006). Unconscious processing of Arabic numerals in unilateral neglect. *Neuropsychologia, 44*, 1999–2006.

Cavézian, C., Rossetti, Y., Danckert, J., d'Amato, T., Dalery, J., & Saoud, M. (2007). Exaggerated leftward bias in the mental number line of patients with schizophrenia. *Brain and Cognition, 63*, 85–90.

Chochon, F., Cohen, L., van de Moortele, P. F., & Dehaene, S. (1999). Differential contributions of the left and right inferior parietal lobules to number processing. *Journal of Cognitive Neuroscience, 11*, 617–630.

Cipolotti, L., Butterworth, B., & Denes, G. (1991). A specific deficit for numbers in a case of dense acalculia. *Brain, 114*, 2619–2637.

Cipolotti, L., Warrington, E. K., & Butterworth, B. (1995). Selective impairment in manipulating Arabic numerals. *Cortex, 31*, 73–86.

Clark, J. M., & Campbell, J. I. D. (1991). Integrated versus modular theories of number skills and acalculia. *Brain and Cognition, 17*, 204–239.

Cohen, L., & Dehaene, S. (1996). Cerebral networks for number processing. Evidence from a case of posterior callosal lesion. *Neurocase, 2*, 155–174.

Cohen, L., Dehaene, S., & Verstichel, P. (1994). Number words and number nonwords: A case of deep dyslexia extending to arabic numerals. *Brain, 117*, 267–279.

Cohn, R. (1971). Arithemtic and learning disabilities. In H. Myklebust (Ed.), *Progress in learning disabilities*. New York: Grune & Stratton.

Collington, R., LeClerq, C., & Mahy, J. (1977). Etude de la semologie des troubles du calcul observes au cours des lesions corticales. *Acta Neurologica Belgica, 77*, 257–275.

Dagenbach, D., & McCloskey, M. (1992). The organization of numberfacts in memory: Evidence from a brain-damaged patient. *Brain and Cognition, 20*, 345–366.

De Hevia, M. D., Girelli, L., & Vallar, G. (2006). Numbers and space: A cognitive illusion? *Experimental Brain Research, 168*, 254–264.

Dehaene, S. (1992). Varieties of numerical abilities. *Cognition, 44*, 1–42.

Dehaene, S. (2003). The neural basis of the weber-fechner law: A logarithmic mental number line. *Trends in Cognitive Sciences, 7*(4), 145–147.

Dehaene, S. (2011). *The number sense. How the mind creates mathematics. Revised and updated edition*. New York: Oxford University Press.

Dehaene, S., Bossini, S., & Giraux, P. (1993). The mental representation of parity and numerical magnitude. *Journal of Experimental Psychology: General, 122*, 371–396.

Dehaene, S., & Cohen, L. (1991). Two mental caculation systems: A case study of severe acalculia with preserved approximation. *Neuropsychologia, 29*(11), 1045–1074.

Dehaene, S., & Cohen, L. (1995). Towards an anatomical and functional model of number processing. *Mathematical Cognition, 1,* 83–120.

Dehaene, S., & Cohen, L. (1997). Cerebral pathways for calculation: Double dissociations between gerstmann's acalculia and subcortical acalculia. *Cortex, 33,* 219–250.

Dehaene, S., Naccache, L., Clec'H, G. L., Koechlin, E., Mueller, M., Dehaene- Lambertz, G., van de Moortele, P. F., & Bihan, D. L. (1998). Imaging unconscious priming. *Nature, 395*(6702), 597–600.

Dehaene, S., Piazza, M., Pinel, P., & Cohen, L. (2003). Three parietal circuits for number processing. *Cognitive Neuropsychology, 20,* 487–506.

Dehaene, S., Spelke, E. S., Pinel, P., Stanescu, R., & Tsivkin, S. (1999). Sources of mathematical thinking: Behavioral and brain-imaging evidence. *Science, 284,* 970–974.

Delazer, M., & Benke, T. (1997). Arithmetic facts without meaning. *Cortex, 33,* 697–710.

Delazer, M., Karner, E., Zamarian, L., Donnemiller, E., & Benke, T. (2005). Number processing in posterior cortical atrophy—A neuropsychological case study. *Neuropsychologia, 44,* 36–51.

Eger, E., Sterzer, P., Russ, M. O., Giraud, A.-L., & Kleinschmidt, A. (2003). A supramodal number representation in human intraparietal cortex. *Neuron, 37*(4), 719–725.

Fayol, M., Barrouillet, P., & Marinthe, C. (1998). Predicting arithmetic achievement from neuropsychological performance: A longitudinal study. *Cognition, 68,* 63–70.

Ferro, J. M., & Botelho, M. A. S. (1980). Alexia for arithmetical signs: A cause of disturbed calculation. *Cortex, 16,* 175–180.

Fischer, M. H. (2001). Number processing induces spatial performance biases. *Neurology, 57,* 822–826.

Fisher, M. H., Castel, A. D., Dodd, M. D., & Pratt, J. (2003). Perceiving numbers causes spatial shifts of attention. *Nature Neuroscience, 6,* 555–556.

Gallistel, C. R., & Gelman, R. (2000). Non-verbal numerical cognition: From reals to integers. *Trends in Cognitive Sciences, 4,* 59–65.

Galton, F. (1880). Visualised numerals. *Nature, 21,* 252–256.

Gazzaniga, M. S., & Hillyard, S. A. (1971). Language and speech capacity in the right hemisphere. *Neuropsychologia, 9,* 273–280.

Gazzaniga, M. S., & Smylie, C. E. (1984). Dissociation of language and cognition: A psychological profile of two disconnected hemispheres. *Brain, 107,* 145–153.

Geary, D. C., Hamson, C. O., & Hoard, M. K. (2000). Numerical and arithmetical cognition: A longitudinal study of process and concept deficits in children with learning disability. *Journal of Experimental Child Psychology, 77,* 236–263.

Gevers, W., Reynvoet, B., & Fias, W. (2003). The mental representation of ordinal sequences is spatially organized. *Cognition, 87,* 87–95.

Ghatan, P. H., Hsieh, J. C., Petersson, K. M., Stone-Elander, S., & Ingvar, M. (1998). Coexistence of attention-based facilitation and inhibition in the human cortex. *NeuroImage, 7,* 23–29.

Gibbon, J. (1977). Scalar expectancy theory and weber's law in animal timing. *Psychological Review, 84,* 279–335.

Goldstein, K. (1948). *Language and language disturbances.* New York: Grune and Stratton.

Grafman, J. (1988). Acalculia. In F. Boller, J. Grafman, G. Rizzolatti, & H. Goodglass (Eds.), *Handbook of neuropsychology* (Vol. 1, pp. 121–136). Amsterdam: Elsevier.

Grafman, J., Kampen, D., Rosenberg, J., Salazar, A. M., & Boller, F. (1989). The progressive breakdown of number processing and calculation ability: A case study. *Cortex, 25,* 121–133.

Hecaen, H., Angelerges, T., & Houllier, S. (1961). Les varietes cliniques des acalculies au cours des lesions retrorolandiques. *Reviews Neurology, 105,* 85–103.

Henschen, S. E. (1925). Clinical and anatomical contributions on brain pathology. *Archives of neurology and psychiatry, 13,* 226–249.

Jewell, G., & McCourt, M. E. (2000). Pseudoneglect: A review and meta-analysis of performance factors in line bisection tasks. *Neuropsychologia, 38,* 93–110.

Johnson, P. J., & Myklebust, H. R. (1967). *Learning disabilities.* New York: Grune & Stratton.

Jong, B. M. D., van Zomeren, A. H., Willemsen, H. T. M., & Paans, A. M. J. (1996). Brain activity related to serial cognitive performance resembles circuitry of higher order motor control. *Experimental Brain Research, 109,* 136–140.

Karmiloff-Smith, A. (1998). Development itself is the key to understanding development disorders. *Trends in Cognitive Sciences, 2,* 389–398.

Laing, E., Hulme, C., Grant, J., & Karmiloff-Smith, A. (2001). Learning to read in Williams Syndrome: Looking beneath the surface of atypical reading development. *Journal of Child Psychology and Psychiatry, 42,* 729–739.

Lee, K. M. (2000). Cortical areas differentially involved in multiplication and subtraction: A functional magnetic resonance imaging study and correlation with a case of selective acalculia. *Annals of Neurology, 48,* 657–661.

Lefevre, J., Bisanz, J., Daley, K. E., Buffone, L., Greenham, S. L., & Sadesky, G. S. (1996). Multiple routes to solution of single-digit multiplication problems. *Journal of Experimental Psychology: General, 3,* 284–306.

Lemer, C., Dehaene, S., Spelke, E. S., & Cohen, L. (2003). Approximate quantities and exact number words: Dissociable systems. *Neuropsychologia, 41,* 1942–1958.

Lewandowsky, M., & Stadelmann, E. (1908). Über einen bemerkenswerten Fall von Hirnblutung und über Rechenstörungen bei Herderkrankung des Gehirns. *Journal für Psychologie und Neurologie, 11,* 249–265.

Loftus, A. M., Nicholls, M. E. R., Mattingley, J. B., & Bradshaw, J. L. (2008). Left to right: Representational biases for numbers and the effect of visuomotor adaptation. *Cognition, 3,* 1048–1058.

Luria, A. R. (1973). *The working brain.* New York: Basic Books.

Luria, A. R. (1976). *Basic problems in neurolinguistics.* New York: Mouton.

McCloskey, M., Aliminosa, D., & Sokol, S. M. (1991). Facts, rules, and procedures in normal calculation: Evidence from multiple single-patient studies of impaired arithmetic facts retrieval. *Brain and Cognition, 17,* 154–203.

McCloskey, M., Caramazza, A., & Basili, A. (1985). Cognitive mechanisms in number processing and calculation: Evidence from dyscalculia. *Brain and Cognition, 4,* 171–196.

Menon, V., Rivera, S. M., White, C. D., Glover, G. H., & Reiss, A. L. (2000). Dissociating prefrontal and parietal cortex activation during arithmetic processing. *NeuroImage, 12,* 357–365.

Molko, N., Cachia, A., Rivière, D., Mangin, J.-F., Bruandet, M., Bihan, D. L., et al. (2003). Functional and structural alterations of the intraparietal sulcus in a developmental dyscaculia of genetic origin. *Neuron, 40*(4), 847–858.

Naccache, L., & Dehaene, S. (2001). The priming methode: Imaging unconscious repetition priming reveals an abstract representation of number in parietal lobes. *Cerebral Cortex, 11,* 966–974.

Paterson, S. J., Brown, J. H., Gsodl, M. K., Johnson, M. H., & Karmiloff-Smith, A. (1999). Cognitive modularity and genetic disorders. *Science, 286,* 2355–2358.

Paterson, S. (2001). Language and number in down syndrome: The complex developmental trajectory from infancy to adulthood. *Down Syndrome Research and Pratice, 7,* 79–86.

Paterson, S. J., Girelli, L., Butterworth, B., & Karmiloff-Smith, A. (2006). Are numerical impairments syndrome specific? Evidence from Williams syndrome and down's syndrome. *Journal of Child Psychology and Psychiatry, 47,* 190–204.

Pesenti, M., Seron, X., & van den Linden, M. (1994). Selective impairment as evidence for mental organisation of number facts: Bb, a case of preserved subtraction? *Cortex, 30,* 661–671.

Pesenti, M., Thioux, M., Seron, X., & de Volder, A. (2000). Neuroanatomical substrates of arabic number processing, numerical comparison and simple addition: A pet study. *Journal of Cognitive Neuroscience, 12,* 461–479.

Piazza, M., Izard, V., Pinel, P., Bihan, D. L., & Dehaene, S. (2004). Tuning curves for approximate numerosity in the human intraparietal sulcus. *Neuron, 44*(3), 547–555.

Piazza, M., Mechelli, A., Butterworth, B., & Price, C. J. (2002). Are subitizing and counting implemented as separate or functionally overlapping processes? *NeuroImage, 15,* 435–446.

Pinel, P., Dehaene, S., Rivière, D., & Bihan, D. L. (2001). Modulation of parietal activation by semantic distance in a number comparison task. *Neuroimage, 14,* 1013–1026.

Pinel, P., Piazza, M., Bihan, D. L., & Dehaene, S. (2004). Distributed and overlapping cerebral representations of number, size, and luminance during comparative judgments. *Neuron, 41,* 983–993.

Priftis, K., Zorzi, M., Meneghello, F., Marenzi, R., & Umiltà, C. (2006). Explicit versus implicit processing of representational space in neglect: Dissociations in accessing the mental number line. *Journal of Cognitive Neuroscience, 18,* 680–688.

Rivera, S. M., Menon, V., White, C. D., Glaser, B., & Reiss, A. L. (2002). Functional brain activation during arithmetic processing in females with fragile x syndrome is related to fmr1 protein expression. *Human Brain Mapping, 16,* 206–218.

Roland, P. E., & Friberg, L. (1985). Localization of cortical areas activated by thinking. *Journal of Neurophysiology, 53,* 1219–1243.

Rosselli, M., & Ardila, A. (1989). Calculation deficits in patients with right and left hemisphere damage. *Neuropsychologia, 27,* 607–618.

Rossetti, Y., Jacquin-Courtois, S., Rode, G., Ota, H., Michel, C., & Boisson, D. (2004). Is action the link between number and space representation? Visuo-manual adaptation improves number bisection in unilateral neglect. *Psychological Science, 15,* 426–430.

Rueckert, L., Lange, N., Partiot, A., Appollonio, I., Litvar, I., Bihan, D. L., et al. (1996). Visualizing cortical activation during mental calculation with funtional mri. *NeuroImage, 3,* 97–103.

Rusconi, E., Kwan, B., Giordano, B. L., Umiltà, C., & Butterworth, B. (2006). Spatial representation of pitch height: The SMARC effect. *Cognition, 99,* 113–129.

Seron, X., Pesenti, M., Noël, M.-P., Deloche, G., & Cornet, J. A. (1992). Images of numbers, or when 98 is upper left and 6 sky blue. *Cognition, 44,* 159–196.

Seymour, S. E., Reuter-Lorenz, P. A., & Gazzaniga, M. S. (1994). The disconnection syndrome: Basic findings confirmed. *Brain, 117,* 105–115.

Simon, O., Mangin, J. F., Cohen, L., Bihan, D. L., & Dehaene, S. (2002). Topographical layout of eye, calculation, and language-related areas in the human parietal lobe. *Neuron, 33,* 475–487.

Stanescu-Cosson, R., Pinel, P., van de Moortele, P., Bihan, D. L., Cohen, L., & Dehaene, S. (2000). Understanding dissociations in dyscalculia: A brain imaging study of the impact of number size calculation. *Brain, 123,* 2240–2255.

Temple, C. (1991). Procedural dyscalculia and number fact dyscalculia: Double dissociation in developmental dyscalculia. *Cognitive Neuropsychology, 8,* 155–176.

Temple, C. M., & Carney, R. (1996). Reading skills in children with Turner's syndrome: An analysis of hyperplexia. *Cortex, 32,* 335–345.

Temple, C. M., & Marriot, A. J. (1998). Arithmetical Ability and Disability in Turner's syndrome: a cognitive neuropsychological analysis. *Developmental Neuropsychology, 14,* 47–67.

Turner, H. (1938). A syndrome of infantilism, congenital webbed neck and cubitus valgus. *Endocrinology, 28,* 566–574.

van Harskamp, N. J., Rudge, P., & Cipolotti, L. (2002). Are multiplication facts implemented by the left supramarginal and angular gyri? *Neuropsychologia, 40,* 1786–1793.

Walsh, V. (2003). A theory of magnitude: Common cortical metrics of time, space and quantity. *Trends in Cognitive Science, 7*(11), 483–488.

Warrington, E. K. (1982). The fractionation of arithmetical skills: A single case study. *Quaterly Journal of Experimental Psychology, 34,* 31–51.

Zamarian, L., Egger, C., & Delazer, M. (2007). The mental representation of ordered sequences in visual neglect. *Cortex, 43,* 542–550.

Zorzi, M., Priftis, K., & Umilta, C. (2002). Neglects disrupts the mental number line. *Nature, 417,* 138–139.

Part II
The Transition from System 1 to System 2

Chapter 5
Possible Explanations

5.1 Subitizing

The first part of this book emphasised several experiments proving how humans resort to two systems of numerical representation: one is inborn and approximative, while the other is culture-influenced, language-dependent, and lies at the basis of exact knowledge. Nevertheless, it is still unclear how these two systems interact with each other to provide an accurate representation of natural numbers. In other words, the seminal question of whether mathematics has developed starting from the approximative system or rather from the precise system remains unanswered.

In some of his works, Stanislas Dehaene supported the idea that both systems are necessary to develop precise arithmetic computation, but that the approximative system is more rudimental, because it contains the basic components of the concept of number. For example, in a work written in partnership with Feigenson, he observed that: "two distinct core systems of numerical representations are present in human infants and in other animal species (...) These systems account for our basic numerical intuitions, and serve as the foundation for the more sophisticated numerical concepts that are uniquely human" (Feigenson et al. 2004, p. 307).

More recently, the same idea was presented in the following passage:

> The linguistic and core-knowledge hypotheses are not necessarily mutually exclusive. Linguistic symbols may play a role, possibly transiently, in the scaffolding process by which core systems are orchestrated and integrated (10, 15). Furthermore, mathematics encompasses multiple domains, and it seems possible that only some of them may depend on language. For instance, geometry and topology arguably call primarily upon visuospatial skills whereas algebra, with its nested structures akin to natural language syntax, might putatively build upon language skills. (Amalric and Dehaene 2016, p. 1)

Nevertheless, in other works, Dehaene contended that the approximative quantity system is the only basic mathematical system, i.e. the one on which all human mathematical representations are built, since it provides humans with the

© The Author(s) 2018
M. Graziano, *Dual-Process Theories of Numerical Cognition*,
SpringerBriefs in Philosophy, https://doi.org/10.1007/978-3-319-96797-4_5

basic vehicle needed to learn the use of numerical symbols. When writing about the accumulator, he states that the approximative system is the fundamental one, because it lies at the basis of our arithmetical abilities (Molko et al. 2004, p. 45). Assuming that this is true, what is the relationship between the accumulator and the development of mathematical knowledge? In other words, which cognitive resources allow humans to overcome the genetically codified approximation mechanisms of their brains and learn the strict rules of exact arithmetic?

In the search for an answer to this question, in his book *The Number Sense*, Dehaene lists a long series of experiments aimed at showing how our computational skills use different resources to provide a representation of the first three positive integers; one, two, three.

The author contends that humans do not count numbers up to three, but that they immediately perceive their presence because their brains identify these quantities effortlessly and without the use of computational resources (Dehaene 1997). This conclusion comes from the experiments carried out by Mandler and Shebo (1982), where subjects had to determine as fast as possible the quantities related to some items presented on a monitor. The data collected showed that reaction times increased linearly by approximately 300 ms only for the items with a numerosity comprised between 4 and 6, while for the items related to a numerosity between 1 and 3 reaction times were very short. On the other hand, when the numerosity exceeded 7, reaction times were more or less constant, while the accuracy of answers dropped. According to the authors, these results indicated that, when faced with small numerosities (from 1 to 3 or 1 to 4), humans do not count items one by one, but they immediately recognise the numerosity; their perception of quantity is instantaneous. For numbers from 4 to 6, reaction times constantly increased by 300 ms; the time needed to process (and therefore count) each and every number. For numbers bigger than 7, reaction times kept unchanged, while the error rate spiked (Fig. 5.1).

The technical term used by the authors to identify this process is *subitizing*, a term coined from the Latin word *subitus* and used to indicate the swift and accurate processing of numerosity in the case of sets composed of maximum 6 items.

Even today, subitizing is a disputed and contested process.

There are authors, such as Mandler and Shebo, who believe that it is the immediate perception of spatial configurations that allows for the representation of 1 as a point, 2 as a line, 3 as a triangular shape, while 4 is subitized only when it is

Fig. 5.1 Reaction times for different numbers of items (Mandler and Shebo 1982)

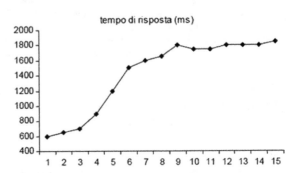

visualised as a usual configuration, such as a square or a triangle with a point at its centre. Beyond 4, the variability of configurations increases, making an immediate recognition completely impossible (Mandler and Shebo 1982).

To corroborate this hypothesis, the results of the study of Trick and Pylyshyn (1993) proved that subitizing is stronger when items are clearly distinguishable from the background, while it is weaker when distinguishing the items requires a lot of attention, such as when counting letters "O" in a group composed of "Q"s. Trick and Pylyshyn also provided a different explanation to the causes of subitizing. According to them, subitizing and enumeration are two consequences of the way in which the human visual system is built. When processing a scene, the visual system works in two stages: a parallel stage (pre-attentive processing) and a periodic stage (attentive processing). Furthermore, they highlighted that the spatial distribution of items affects subitizing only partially, on the condition that the items are clearly identifiable (Atkinson et al. 1976), contrary to what happens with sets of items that have to be counted. The authors inferred that subitizing depends on pre-attentive processing, while enumeration requires attentive processing. During pre-attentive processing, spatial markers called FINSTs (*FINgers of INSTanciation*) are associated to items. The term subitizing, hence, is structural and corresponds to the number of FINSTs that can be associated to different entities in the visual field. To establish the cardinality of a set of items, subjects simply evaluate the number of activated FINSTs, without paying attention to the items (Pylyshyn 1998).

When the number of items exceeds the number of FINSTs available or when the positions of the items are not clearly distinguishable, the subjects resort to enumeration. Adults have only 4 FINSTs at their disposal and therefore subitizing is limited to sets of a maximum of 4 items. Unfortunately, the authors do not explain why adults do not have 5 or 8 FINSTs, but they simply suggest that the number of FINSTs might increase with age and vary according to every single individual, despite the fact that there are no studies supporting this hypothesis.

Gallistel and Gelman (1992) put forward an even more radical explanation, contending that subitizing is a very swift computation based on non-verbal markers, i.e. an inborn and pre-verbal counting skill. The authors hypothesised that humans have an enumeration system similar to the one used by animals, which is very fast, but inaccurate. This conclusion—that subitizing is indeed an enumeration process —comes from the variable reaction times in the cases of numbers from 1 to 4, but also from the sharp increase in reaction times when the items went from 1 to 2 and from 2 to 3. Furthermore, features such as the regularity of the sets and concentricity do influence subitizing, showing that the presence of subitizable groups depends on the physical features of the set. According to the arrangement of items, a small set can be perceived as a single group and therefore lead to one attention focus, or it can be perceived as of composed of several groups and therefore different subgroups. As a matter of fact, when items are arranged irregularly, different sub-groups emerge because of the Gestalt proximity principle, which identifies the items nearer to each other as belonging to the same group. On the contrary, when items are arranged regularly, no sub-groups are identified, moving the attention focus to the whole group.

Fig. 5.2 Examples of 4-point
sets irregularly and regularly
arranged

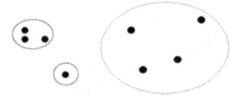

Raphaëlle Lépine and colleagues (Lépine et al. 2003) proved that adults and
children of 10 years of age faced with sets containing 1 to 8 points with changing
arrangements had slower reaction times when faced with irregular sets (Fig. 5.2).

Nevertheless, it is important to highlight that for sets with a subitizable size,
regularity had a stronger effect on those containing 4 items compared to those with
3 items (for obvious reasons, the regularity principle does not apply in sets with 1 or
2 items). Furthermore, items had to be located in clearly distinguishable positions to
be swiftly subitizable.

The fact that the distinguishing process differs between sets with small and big
numerosities from a qualitative perspective was also proved by recent psychological
and physical experiments (cfr. Revkin et al 2008; Piazza et al 2011). By taking into
account the outcomes of these studies, it is possible to contend that subitizing small
numbers is a process based on a cognitive mechanism that is different to the simple
estimation of other quantities (accumulator) and, therefore, that small numbers are
not represented as approximate numerical quantities.

Nevertheless, contrary to what Dehaene assumed at the beginning of his research
on numerical cognition, subitizing is not an effortlessly "pre-attentive" process, but
rather a process requiring attention (Dehaene 2011). As the author admits in the last
part of the second edition of *Number Sense*:

> I must confess, however, that there is one point where I got it wrong. [..]I described "subitizing,"
> or the remarkable capacity that we all have to iden- tify 1, 2, or 3 items at a glance. I was correct
> in suggesting that we all can "subitize" with- out counting—a whole stream of novel publi-
> cations has confirmed this point with a variety of methods. However, I was wrong in suggesting
> that subitizing is essentially a form of "precise approximation." (Dehaene 2011, pp. 256–257)

Further on:

> How subitizing actually works remains something of a mystery. One interesting clue,
> however, is that contrary to what we once thought, it is not independent of our attention.
> Subjectively, subitizing seems to be automatic: One glance at a set seems enough to effort-
> lessly recognize that it contains 1, 2 or 3 objects. This is an illusion, however. [..]Far from
> being "pre-attentive" and effortless, subitizing requires attention. We can select a small
> number of items, and even track them through time and space, but this taxes our attention. So
> how does subitizing work? Current research suggests that we have 3 or 4 memory slots
> where we can temporarily stock a pointer to virtually any mental representation. This
> memory store is called "working memory"—a transient supply that keeps the objects of
> thought on-line for a brief moment. [..] There is nothing in the approximate number system
> to support a system of exact arithmetic with discrete numbers. (Dehaene 2011, pp. 259–260)

In conclusion, there is still some controversy around the interpretations put
forward by several authors. First of all, the fact that subitizing depends on the

numerosity of the items or, on the contrary, on other factors related to perception, such as the area covered by the items or the duration of the stimuli. According to Feigenson, Carey and Spelke (2002), the general configuration of the stimulus affects the process of subitizing more than the number of items shown. Therefore, it is still unclear whether the majority of empirical evidence linked to subitizing can be better explained by perception factors that are not related to numerical cognition. Secondly, the attention issue cannot be overlooked. The mere fact that subitizing requires "attention" entails that its codification principle is radically different to the way in which numbers are codified on an approximative numerical line. The distinguishing feature of subitizing is that it provides a "discrete" estimation for each number between 1 and 3, while the approximative system cannot provide a discrete arithmetic system with accurate numbers.

It is worth noting, though, that attention and accuracy seem to be features belonging to system 2, where mathematically literate humans spontaneously start counting numerosities and where symbolic representations play a pivotal role. This topic will be tackled in the next paragraph.

5.2 Conceptual and Procedural Knowledge

Several authors (cfr. Butterworth 1999; Geary 1994; Gelman 1993) have repeatedly put forward the idea that arithmetic computation represents a "privileged" discipline, i.e. a domain where evolution favoured some skills, in that they became particularly easy to acquire and early developable. For example, children can understand at an early age the qualitative effects of addition and subtraction, even before mastering proper arithmetic procedures.

Nevertheless, this idea has its opponents, who tend to prefer the "exposition frequency" hypothesis, which states that some skills are developed early because the environment provides a large number of opportunities of observation and imitation. For example, Fuson (1988) put forward the idea that children infer the concepts at the basis of proper procedures from their counting activities and the observation of others. Contrary to the privileged domain hypothesis, the exposition frequency hypothesis contends that even before developing a proper conceptual knowledge, children develop procedural skills in the domains that provide sufficient opportunities of observation and imitation (i.e. simple actions aimed at solving selected types of problems).

It goes without saying that the issue of the relationship between conceptual and procedural skills is a particularly interesting one. Furthermore, a deeper look into the interactions between these two types of knowledge could contribute to a better understanding of the mechanisms underlying numerical cognition. It is not by chance that several researchers have tried to investigate the relationships between conceptual and procedural knowledge by focussing on some arithmetic operations such as simple additions, and additions and subtractions with multiple digits.

As far as simple additions go, the skills developed by children require the understanding that each number is represented once and only once, and that the order of the addition does not affect the final result (commutative property). Several studies investigated the relationship between this conceptual knowledge and the use of the "min" strategy based on the commutative property. For example, Baroody and Gannon (1984) analysed this relationship in nursery school children.

To evaluate the understanding of the commutative property, the authors presented children with some fast computation tasks such as $4 + 2$ and $2 + 4$. The children were asked if the two operations led to the same result. In another variations of this task, after having children answer the addition $6 + 4$, they were asked whether $4 + 6$ gave the same result or not. From this experiment, the authors came to the conclusion that children can understand the commutative property even before using the "min" strategy. In another study, Siegler and Crowley (1994) asked 5-year-old children to use procedures in line with or going against the principles of addition (the sum strategy, the "min" strategy and a wrong procedure, where one of the two operands was provided twice). Children that had never used the "min" procedure correctly identified the procedure as right (as the sum), but rejected the wrong procedure. As in the previous example, these studies led the authors to the conclusion that children understand the principles at the basis of addition even before using its procedures.

Another area that provided fruitful results for the research on the relationship between conceptual and procedural knowledge is the study on quantification processes. According to Halford (1993), it is impossible to develop the concept of number without quantification; quantification being the procedure that allows for the assignment of numerical values to sets, the evaluation of the effects of format in different sets or the understanding of the complex relationship existing among numbers. Therefore, according to the author, quantification plays a basic role in the identification of the numerosity of a set of items.

Among enumeration processes, researchers focussed mainly on the one that has often been considered at the basis of all other arithmetic computation tasks. Enumeration is indeed a verification test that empirically checks the validity of a reasoning process, for example when conservation tasks are involved or in the case of arithmetic tasks (Groen and Parkman 1972; Svenson 1975). It is important to highlight that the majority of researchers believe in the existence of a natural inborn tendency to identify discrete quantities (Briars and Siegler 1984; Fuson 1988; Gallistel and Gelman 1992; Gelman and Gallistel 1978; Resnick 1986; Wynn 1990), even though not all of them ascribe the same importance to inborn and practical skills. As far as enumeration in childhood is concerned, two opposed schools of thought can be identified: the "principle-first" theory, and the "principle-later" theory.

The "principle-later" theory states that principles are progressively (and gradually) extrapolated from the repeated practice of enumeration, which is acquired through imitation (Briars and Siegler 1984; Fuson 1988; Fuson and Hall 1983). Initially, enumeration starts as a purposeless activity, a routine, while over time children discover its bonds with cardinality (Fuson 1988; Wynn 1990). This concept does not go against the idea that children have an inborn sensitivity towards numbers that provides the basis on which arithmetic learning is built, but it does not account for its backbone. On the contrary, the "principle-first" theory states that the principles that govern enumeration are inborn and therefore that children master them even before starting enumerating, allowing them to recognise enumeration as a sensible operation that can be acquired and controlled. Gelman and Gallistel (1978) identified 5 principles:

(1) the one-one principle: when counting items in a set, each item is associate to one and only one number name;
(2) the stable-order principle: the list of number names is an ordered list, a fixed sequence;
(3) the cardinal principle: the last number name used equals the total number of items in the set;
(4) the abstraction principle: the heterogeneity (vs. the homogeneity) of the items in the set does not affect the counting procedure;
(5) the order-irrelevance principle: the order in which the items in the set are counted does not affect the total number of items in the set.

According to this model, children have a tacit knowledge of counting principles, similar to the tacit knowledge of the grammar of a language (Chomsky 1957). This knowledge allows children to recognise the procedures that are correct, exactly as the knowledge of grammar allows them to recognise the correct sentences in a language, even if they have never heard them (Gelman and Greeno 1989). Furthermore, the authors highlighted that another parallel can be drawn between linguistic and mathematical knowledge: exactly as grammar provides the possibility to produce an infinite number of new sentences (generative grammar), the knowledge of principles allows for counting strategies adaptable to different tasks. Hence, when counting with number names, it is necessary to use them in an ordered list (stable-order principle): according to this definition, any list (even "one", "two", "six", "three", "eight") can lead to a successful result. Furthermore, a successful result does not require the use of number names: the labels used can be non-verbal (e.g. fingers or different parts of the body that represent numerosities, the basis of counting). Therefore, according to Gallistel and Gelman (1992) the skills needed for counting are not linguistic and they are accessible to both animals and children.

All principles identified by Gelman and Gallistel have been tested on children. Gelman and Meck (1983) have studied how 3–4-year-old children understand the "one-one principle" and the "order-irrelevance principle" by asking them to evaluate the counting procedures carried out by a puppet. These authors focussed in particular on the tests in which the puppet counted the items correctly, but without following a standard order from the left to the right. Even if several children had never witnessed a counting procedure, the majority of them estimated that it was correct, showing that they were ready to accept starting counting from any item and not necessarily from the one placed further on the left.

These studies suggest that the order-irrelevance principle is accessible to children at an early age. Furthermore, even children that have not been initiated to verbal numerical chains tend to use a number of labels that equals the number of items that have to be counted (for example, when faced with two-item sets, children could say "two, five"), in line with the one-one principle (Gelman and Gallistel 1978). These studies were strongly criticised. Briars and Siegler (1984) did not achieve the same results of Gelman and Meck in 3-year-old children, but only in 5-year-old children. Gelman and Meck (1986) justified this difference by saying that in Briars and Siegler's study, children had to evaluate how conventional the counting procedure was and not how accurate it was.

A study carried out by Baroody (1984) highlighted another issue with the early understanding of the order-irrelevance principle, showing that children believe that counting in different orders can lead to different results. Nevertheless, Gelman, Meck and Herking (Gelman et al 1986) replied by saying that in Baroody's experiment, children gave two different results because they considered the question of the researcher (e.g. "if we change the order, which result do we get?") as an indicator of the fact that their first answer (the result of counting) was wrong.

Another principle identified by Gelman and colleagues that was harshly commented and criticised is the cardinal principle. According to Fuson and Hall (1983), the fact that in Gelman's experiment children repeated the last number name used and understood the eventual mistake made by the puppet was due to simple imitation. Fuson and Hall carried out an experiment in which they asked children to count the items in a set and then asked them "how many items are there?". Their study showed that children did not answer with a number name, but they rather started counting again. Furthermore, their answers were different if the question was related to a specific class of items ("how many flowers are there?") and they did not spontaneously count the items when they had to provide a precise answer. Nevertheless, other experiments carried out by different researchers did not come to the same outcomes.

Finally, as far as the abstraction principle is concerned, several experiments showed that children can count the animate or inanimate items of a heterogeneous set already at an early age (Fuson, Pergament and Lyons 1985; Gelman and Tucker 1975), even in the case of actions or sounds (Wynn 1990). These experiments showed that children own an abstraction principle, despite the fact that Shipley and Shepperson contested this conclusion (Shipley and Shepperson 1990). The two researchers tried to prove that counting performance depended on the physical features of items. For example, children often encountered difficulties when counting different types of items or different colours in the same set. Nevertheless, the complexity level of counting required by these experiments seemed to go well beyond the simple act of counting: as a matter of fact, counting the items belonging to a category requires a higher level of abstraction (e.g. being able to distinguish categories and rank items in different categories) compared to the simple act of counting items.

Therefore, one question remains unanswered: how can children initially endowed with a non-verbal system of quantity representation eventually acquire the concept of integer numbers? Susan Carey (2001) tried to answer this important question by developing a theory called "bootstrapping".

5.2.1 The Theory of Bootstrapping

The theory of bootstrapping developed by American psychologist Susan Carey suggests a new stance on cognitive development that has the potential of overcoming the constraints found in Piaget's theory.

The first part of this book explained how Piaget's multi-stage theory postulated the existence of broad and pervasive cognitive structures that allowed the subject to process different contents (i.e. "generics for domains"), hence leading to a categorisation of children based on their development stages. Nevertheless, after a long series of experiments carried out both by development psychologists and experts of numerical cognition (cfr. Dehaene 2011), it was concluded that reasoning abilities and skills in children are subject to a higher level of variability and heterogeneity than what was envisaged by Piaget's theory.

Carey's approach stands in direct opposition to Piaget's idea that there are general cognitive structures and general changes for each domain, while supporting the idea of "domain-specific" cognitive structures and changes that deal with different contents. With her approach, the researcher shifts the focus from intellectual operations—which are pivotal in Piaget's theory—to the content of reasoning, i.e. the concept and conceptual networks in which different types of knowledge are framed (Carey 2009a).

A question arises: how do we learn concepts?

According to Susan Carey "Concepts are individuated on the basis of two kinds of considerations: their reference to different entities in the world and their role in distinct mental systems of inferential relations" (Carey 2004, p. 60).

There are indeed some concepts that are not instantly perceivable (e.g. scientific notions such as quarks or microbes) and that can only be acquired through the vehicle provided by language. Nevertheless, there are concepts that children master even without specific training. For example, Carey cites an experiment carried out by Hespos and Spelke in 2004 that provided data proving the existence of universal concepts (Hespos and Spelke 2004).

The two authors focussed on the comparison of the English and Korean languages: in Korean, two different words are used to identify a *short* or a *long* contact between items, whereas there is only one word for *contact* in English. Korean children seem to effortlessly understand and use this distinction. Hespos and Spelke proved that even 5-month-old babies are sensitive to the difference between short and long contacts, even though they ignore the meaning of the words short and long: after watching the researcher put two items together for a short time, babies showed renewed interest when faced with items that were matched for a longer time, and vice versa. Furthermore, the authors asked a group of English-speaking adults to evaluate the similarity between the two types of contact: contrary to babies, their earliest judgment was that there was no difference between them. Nevertheless, after thinking about it, the English-speaking adults also noticed a difference between short and long contacts: during interviews after the experiment, a good number of them flagged the difference to the researchers.

In this experiment, therefore, the effect of language is that of shifting the subject's attention to the aspects linked to the environment, rather than other aspects. The existence of a similar mechanism was also postulated by the psycholinguist Asifa Majid and colleagues (Majid et al. 2004). In a paper published in 2004, the authors described different mechanisms that could account for the learning process of new concepts that went beyond non-verbal representations. The first mechanism is shown by the ability to learn new words, which rearranges conceptual categories. For example, when one acquires specific knowledge in a specific field, the concepts used become structured, easier to process, and easier to predict. In maths, concepts that are simple and similar are often grouped together under the umbrella of a higher abstract concept: in this way, the concept of group can be applied to various objects, integers, matrixes, or sets of functions.

The second mechanism is linked to the detection of correspondences by children or adults. When learning new words, children notice that different concepts are grouped under the same word, a phenomenon that allows them to compare different situations and detect new regular patterns. Through this mechanism, language shifts the attention of the subject to the comparison of different situations, allowing for the learning of high-level relationships between different concepts and going beyond the scope of pre-existent concepts.

Carey's theory of bootstrapping on number learning is related to a similar kind of explanation mechanism. According to it, children are initially endowed with a non-verbal system of quantity representation and, starting from it, they develop an understanding of numbers and integers. By comparing their representation of small quantities with the first words of the number list (one, two, three…), children learn the principles upon which the list is built and apply them to all numbers in the list. In so doing, they learn that each number in the list corresponds to a specific quantity and that each quantity is created by taking the previous one and adding 1 to it.

Nevertheless, according to the theory, the total level of abstraction is higher than the one described by Majid: as a matter of fact, children analyse the relationships existing between nearby numerosities not because they share a name, but rather because all names of numerosities belong to the same linguistic object (an ordered list of number names). According to Carey, children in this stage develop the concept of exact number that defines numerosities. In this way, the theory of bootstrapping explains how concepts can be developed starting from the basic language of rudimentary representations.

Following this reasoning, language plays an important role in numerical cognition, because it links the different types of non-verbal representation of numbers. Right after birth, children have two systems to represent numerosities, but none perfectly matches integers. The first one is an approximative representation system that allows them to understand big numerosities, despite the fact that they mix up nearby numerosities such as 15 and 16. The second one is a system limited to small numerosities that allows for an exact representation of 1, 2 and 3. At the age of one year, children struggle to link these two systems and therefore, while they are able to compare 2 to 3 and 4 to 8, they are not able to compare 2 to 4 (Xu 2003). In this case, as in bootstrapping, language allows them to overcome the original non-verbal representations. The mere fact of mastering a language requires humans to develop a substantial amount of cognitive resources that are beyond the reach of animals. Without language, many strategies are simply impossible, such as all those requiring a memory of verbal work and, in particular, memory related to numerical cognition, counting, and mental computation tasks.

Three main theses stand at the basis of Susan Carey's knowledge acquisition theory, including the acquisition of number-related knowledge. The first one is "discontinuity", which postulates that children during their development learn new concepts that change their ability to express themselves. As Jacob Beck highlighted:

> while many two year olds can recite a portion of the count list ("One, two, three, ..."), they don't seem to know what the words in the list mean. If asked for n pennies from a pile, or to point to the card with n fish, they will respond with a random number of pennies or point to a random card. Moreover, their failures consist of more than ignorance of language. [...] Carey concludes that two year olds lack the representational resources to think about the integers. Four year olds, by contrast, have those resources; they succeed on the point-to-a-card and give-me-n tasks. When children first memorize the count list, it serves as a mere placeholder structure. It encodes serial order ("three" comes after "two," which comes after "one"), but the nature of that order is not defined for the children. It's as though they were say- ing "eeny, meeny, miny, mo." Nevertheless, Carey maintains that this placeholder structure plays a crucial role in explaining how children acquire integer concepts, and that similar placeholder structures play an essential role in other episodes of concept learning. (Beck 2017, p. 111)

Placeholders play therefore a pivotal role in generating conceptual discontinuities and represent the second basic thesis. According to Carey, without proper cognitive structures for placeholders, it is impossible to develop new networks of concepts. To support this idea, Carey provides the example of the populations living in social environments without a wide numerical vocabulary, where the development of new networks of concepts is impossible because they never become cardinal-principle knowers (Carey 2009b).

The gaps of conceptual discontinuity are filled by the learning process of bootstrapping (third thesis), which is fed by placeholders but has a wider scope thanks to the influence of education and learning. Through the learning process of bootstrapping, children learn numerical concepts such as "three", "seven", and "ten". Therefore, according to Carey:

> children at this stage have learned to use their object file systems to place models stored in long-term memory in one-to-one correspon- dence with objects in the world, and to associate such states of one-to-one correspondence with the first four number words. So they know that there is "one" object when the object is in one-to-one correspondence with a model of a sin- gleton in long-term memory {i}; that there are "two" objects when the objects are in one-to-one correspondence with a model of a pair of individuals in long-term memory {j, k}; and so on, up to four (the upper bound of the object file sys- tem). Carey calls children at this stage "subset-knowers" and calls the system they use "enriched parallel individuation." Finally, by three-and-a-half or four years of age, children assign meanings to the remainder of the terms in their count list. (Beck 2017, p. 111)

Nevertheless, this model has its detractors. One of the first researchers to criticise bootstrapping was Jerry Fodor, who repeatedly voiced his concern about the possibility that Carey's theory did not in fact explain how individuals improved their abilities to express themselves (cfr. Fodor 2008). Fodor contended that the strongest limitation of the bootstrapping theory lies in the "circularity" of this model. For example, according to Fodor, postulating that placeholders are vehicles for learning new contents also means postulating that children already have concepts with those exact contents. It is obvious that a conceptual domain (target domain) can be enriched starting from a familiar domain (source domain), but by doing so it is necessary to postulate that individuals already possess the concepts of a familiar domain.

For example, the concept of "electric wires" could be understood starting from the familiar domain "hydraulic system", but only on the condition that we have the conceptual resources to specify the properties of hydraulic systems. Fodor believes that Carey's theory features this kind of circular loop, which is also visible in her explanation of how children learn the list of natural numbers. Fodor's claim can be understood even better in an example provided by Rey (2014), who focussed on the role played by the concept of "successor" in the boostrapping theory. Carey writes that it: "supports the induction that any two successive numerals will refer to sets such that the numeral further along the list picks out a set that is one greater than that earlier in the list" (Carey 2009a, p. 477). Nevertheless, Rey contends that: "But here 'is one greater than' expresses the very concept of SUCCESSOR whose acquisition Carey is trying to explain. In the first place, one can ask how this concept even occurs to the child." (Rey 2014, p. 117).

In general, Rey believes that the child must already have the concept SUCCESSOR to entertain the analogy that he takes Carey to credit with generating that concept.

Rey criticised the bootstrapping theory also from another angle, related to the way in which children interpret placeholders. In Carey's theory, children always interpret them in a natural and correct fashion. But why should they? Could children not interpret placeholders unnaturally or improperly?

Similar arguments were formulated by psychologist Lance J. Rips and colleagues, who criticised the bootstrapping theory in a work published in 2008 (Rips et al. 2008) stating that:

"This idea (one word forward [in the count list] equals one more individual) captures the successor principle." Notice, though, that Principle (3) depends on the concept of the next count word, which we have referred to as "s(n)," for any count term "n" (if "n" is "five," "s (n)" is "six"; if "n" is "ninety," "s(n)" is "ninety-one"; etc.). For these purposes, simple counting won't do as a guide to "s(n)," as simple counting uses a finite list of elements. For example, if a child's count list stops at "nine," then Principle (3) can extend the numeral-cardinality connection through nine. In order to capture all the natural numbers, however, Principle (3) requires advanced counting: an appreciation of the full numeral system. But at this point the trouble with the counting hypothesis comes clearly into view, for at the point at which children are supposed to infer Principle (3)—at a little over 4 years of age—they have not yet mas- tered advanced counting. There is nothing that determines for such a number learner which function or sequence specifies the natural number words (i.e., the function that appears as "s(n)" in Principle [3]). (Rips et al. 2008, pp. 631–632)

Further on:

Suppose, for example, that the count system that the child is learning is not one for the natural numbers but, instead, for arithmetic modulo 10, so that adding 1 to 0 produces 1,…, and adding 1 to 8 produces 9, but adding 1 to 9 produces 0, and so on in a cyclical pattern. In this case, Principle (3) is still a valid generalization of (2) if we interpret "s(n)" as the next numeral in the modular cycle, but then what has been learned is not the natural numbers. The generalization in (3) can seduce you if you think of the child as interpreting it (after a year of struggle) as "Aha, I finally get it! The next number in the count sequence denotes the size of sets that have one more thing." But "next number in the count sequence" isn't an innocent expression since the issue is, in part, how children figure out from (2) that the next number is given by the successor function for the numerals corresponding to the natural numbers and not to a different sequence (e.g., the numbers mod 10 or mod 38 or mod 983). (Rips et al. 2008, p. 632)

These extracts show that Rips and colleagues hastily combine the two problems highlighted by Fodor and Rey (circularity and interpretation deviation). Generally speaking, these are two separated issues. In the case of interpretation deviation, it is important to understand why, when children say "seven", they are able to confirm their hypothesis and not the competing ones. In the case of circularity, instead, the core of the issue relates to why, when children say "seven", they refer exactly to seven as an entity.

In any case, the bootstrapping theory does not provide an answer to any of these points. The overall conclusion from the critics moved against Susan Carey's theory is that mathematics is not a domain as any other and, therefore, that it is necessary to develop a theory to explain the transition from knowledge of the object domain to the mathematical domain. According to the American philosopher George Lakoff, this task is performed by "conceptual metaphors", cognitive mechanisms that allow individuals to reason on a set of objects as if they were other objects. In this way, metaphors are not figures of speech, but real reasoning mechanisms. As Lakoff and Núñez write:

"conceptual metaphor" has a technical meaning: It is a grounded, inference-preserving cross-domain mapping-a neural mechanism that allows us to use the inferential structure of one conceptual domain (say, geometry) to reason about another (say, arithmetic). Such conceptual metaphors allow us to apply what we know about one branch of mathematics in order to reason about another branch. (Lakoff and Núñez 2000, p. 6)

Before analysing the way in which Lakoff and Rafael Núñez channelled their ideas in a research paradigm on mathematical concepts through the use of the technical concept of "conceptual metaphor", it is useful and almost necessary to provide an overview of the works carried out by Lakoff in the field of cognitive semantics, the domain were the term "conceptual metaphor" was coined.

5.3 Concepts and Cognitive Semantics

Cognitive psychologists have always considered concepts as "reservoirs" where objects sharing common traits or features are stored, but they also have always wondered how humans are able to identify a "common element" in a heterogeneous group of items. Jerome Bruner (1968) postulated that a concept comes to life when two or more objects or events are grouped or ranked together; these items are then distinguished from the others on the basis of some trait or feature. According to Bruner, concepts depend basically on categorisation activities, i.e. finding similarities in things that may be perceived as different and grouping them in object classes.

The earliest lab studies on the issue assumed the existence of "perception transparency", which suggests that perception directly detects the common traits of different stimuli. Nevertheless, the hypothesis that concepts simply derive from perceivable stimuli was rejected by Ludwig Wittgenstein (1953), who made the example of the word "game", which includes activities and materials that are extremely different, making it difficult to identify perceivable common traits.

Rejecting the idea that concepts are merely lists of characteristics, Eleanor Rosch (1975) put forward the "prototype theory", which postulates that the meaning given to a huge number of words at a cognitive level could be linked to a prototype, i.e. the representation of an archetype. For example, the prototype associated to the word "bird" could be a sparrow rather than a bat, the prototype associated to "tree" could be an oak rather than a palm (even though things would be different for someone born on a tropical island). It takes less time to verify whether a sparrow is a bird than verifying whether a duck is a bird.

In their endeavour to reject the idea that the meanings of linguistic items correspond to independent entities, cognitive psychologists try to describe meanings as cognitive contents. In this way, the study of meanings is embedded in the study of the mental processes that build these contents. The classic division made by traditional linguistics—based on necessary and sufficient conditions—is therefore completely replaced by the use of cognitive structures (conceptual and perceptive) such as prototypes, domains, frames, and—in particular—conceptual metaphors.

The concept of domain was thoroughly studied by Langacker (1987), who focussed in particular on the meaning of the word "Monday". According to the author, "Monday" can be explained only in the wider context of "week" (someone who is not familiar with the concept of a "7-day week" would not understand "Monday"). Furthermore, the concept of week can only be understood in the framework of a recurring cycle of day and night. Langacker defines "7-day week" as the semantic domain that allows for the understanding of the term "Monday", while the "day-night" cycle is the semantic domain used to understand "week".

The rule that applies to lexical terms also applies to morphological and syntactical categories: they as well can be understood in relationship to relevant domains (e.g. the domain to understand the past tense in grammar is time). Time, the tridimensional space and other sensory experiences (e.g. temperature, colour, taste, and tone) are defined by the author as basic domains, concepts that cannot be broken into simpler cognitive structures. Some linguistic labels can simultaneously belong to several different domains; "golf balls", for example, gives an idea in terms of shape, but also of its colour, size, material, etc. Furthermore, there are primary and secondary domains: salt, in its domestic connotation, is associated primarily to food, while only secondarily to the chemical compound. On the other hand, sodium chloride (which refers to the same element) is associated primarily to the chemical compound.

George Lakoff (1987) provided a more complex example of a term that refers simultaneously to several domains: the word "mother". According to the author, the term covers five domains, namely:

(1) the genetic domain: a mother is a female who gives her genetic material to a child;
(2) the birth domain: a mother is a female who gives birth to a child;
(3) the nurturance domain: a mother is a female who nurtures a child;
(4) the genealogical domain: a mother is the closest female ancestor to a child;
(5) the marital domain: a mother is the wife of the father.

By comparing these domains to their homologues for "father":

(1) the genetic domain: a father is a male who gives his genetic material to a child;
(2) the responsibility domain: a father is responsible for the financial wealth of a child (and mother);
(3) the authority domain: a father is responsible for the good behaviour of a child;
(4) the genealogical domain: a father is the closest male ancestor to a child;
(5) the marital domain: a father is the husband of a mother of a child.

Through this comparison, Lakoff comes to the conclusion that the meaning of father does not differ to that of mother just because of the gender trait, but that these terms are only comparable in few limited domains (genetic, genealogical, and marital), while the others only exist in the concept of father (authority and responsibility).

Nevertheless, Langacker himself highlights that his concept of domain largely overlaps with what other researchers have defined as frame, script, scheme, and model. The language used is often confusing, because different authors use different terms to refer to the same concept, while sometimes the same author changes the term in different papers. For example, the concept of frame is often associated to a knowledge network that links multiple domains, while on the other hand script is mostly used to refer to the temporal sequencing and the causal relations that link events and states in certain action frames. Charles Fillmore (1985) had the merit of introducing frames in the debate on cognitive semantics, providing examples of a phenomenon in which words cannot be individually defined, but rather find meaning in correlated sematic fields. The starting assumption of Fillmore was that specific terms, expressions, and grammatical choices are associated to specific frames in our memory and, therefore, the presence of a linguistic structure in an appropriate context triggers a specific frame in the mind of the recipient, giving access to new linguistic material associated to the same frame.

Let us dig into the concept of frame by taking into consideration Lakoff's reasoning on the word "mother". The five domains that specify the word are not a random set of domains, but rather a structured set, a frame related to the word mother. According to this frame, a mother is a female that has sexual intercourse with a father, falls pregnant, gives birth, devotes time to raising a child, and remains married with the father of her child. Clearly, this description reflects only a highly idealised of the concept, since it does not take into account the numerous deviations from this model (e.g. divorce, separation from bed and board, adopted children, single mothers, etc.). In conclusion, frames structure our social and conceptual life. They represent cognitive processes that define words without a clear correspondence to reality. The meaning of "Monday" could not be explained by a theory that combines, for example, the symbol of this word with an entity of reality that represents its content. According to Fillmore, the existence of "Monday" is therefore neither objective nor subjective, but it rather belongs to the reality order defined by the creative ability that all humans share because of their basic structures.

From this perspective, it is clear that the difference between logical theories and the cognitive approach lies in their study object: while logics tackles the relationships between language and reality, cognitive science aims at assessing the mental models that cognitive agents develop after being exposed to specific stimuli, such as linguistic inputs. From the point of view of cognitive science, conceptualisation and —generally speaking—the production and understanding of even the simplest utterance are characterised by the central role of creative processes (prototypes, schemes, scripts, frames, metaphors, etc.). This is due to the fact that our thought mainly works unconsciously, swiftly, and automatically.

Since the majority of knowledge is "hidden", the biggest challenge is to bring it to the surface. According to George Lakoff and Mark Johnson (1980; 1999), the most successful methodology in doing so lays in the use of conceptual metaphors. Human thought is developed starting from "images" that can be visual, auditory, olfactive, gustative, tactile, kinaesthatic, or organic, hence explaining why we often use metaphors and figures of speech when we speak. They help us interpret the inputs coming from the outside world: metaphors deal with imagination, because they are "neural connections made through the co-triggering" of two domains, one belonging to the cerebral areas devoted to the processing of sensomotoric experiences, the other belonging to the cerebral areas that process the subjective experience. Conceptualisation kicks off from these universal metaphors (called "primary metaphors"), leading to the more or less conscious processing of different metaphors (e.g. time is conceptualised through the guiding metaphor that the past is "behind us", the present is "on us", and the future is "ahead of us").

The linguistic "exploitation" of metaphors is only the tip of a massive functional iceberg, where sensory processes and cognitive functions are less divided and independent in their latest stages than what was previously thought. It is not by chance that metaphors—because of their nature of bridge-builders between subjective experience and thought—are the linguistic and conceptual vehicles that best convey this complexity compared to other linguistic phenomena.

Some authors even say that metaphors should be treated as something not only pertaining to the study scope of linguistics.

The American philosopher John Searle (1979) is of this opinion. When dwelling about the meaning of sentences such as "Sally is a block of ice", he said that—if analysed word by word—the sentence was semantically peculiar, since an inanimate item such as ice could not have been predicted from an animate entity such as Sally. The sentence is acceptable only on the condition that the listener/reader goes beyond the literally meaning of words, understanding the metaphoric meaning of the sentence. To do so, listeners/readers must combine their linguistic skills with pragmatic skills. Searle made a distinction between semantic and pragmatic tasks: the former ones deal with literal and purely linguistic meaning, the latter ones deal with the context and the meaning that the sender wishes to convey. Hence, according to Searle's interpretation, understanding metaphoric sentences requires three stages: recognising that the literal meaning (expressed meaning) of the word is insufficient, rejecting this meaning, and looking for a new and different meaning (the intended meaning that makes the sentence meaningful).

David Edward Cooper (1986) strongly criticised Searle's approach by pointing to at least four issues that should be clarified. First of all, the supposed deviance of metaphors implied that all speakers of a certain language could "dismantle" the metaphor in all metaphoric expressions, restoring its full grammatical meaning. Nevertheless, replacing a metaphoric expression with an equivalent that is not a metaphor is often hard or even impossible. Secondly, it is highly counter-intuitive to claim that metaphors, a widely spread linguistic mechanism, can be explained as an exception to the rule. Furthermore, the wide dissemination of metaphors goes against the deviance hypothesis: as an endemic trend, metaphors would become the rule, not the exception. Finally, there is the issue of *bona fide* communicators intentionally producing grammarly deviant sentences only because of the wish of having recipients mobilising all sorts of interpretation abilities to understand their intended meaning.

Since the publishing of *Metaphors we live by* (Lakoff and Johnson 1980) and the development of a cognitive approach to metaphors, metaphors have not been considered as a violation of linguistic rules. The new model postulated that metaphors were a vehicle that made possible the conceptual representation of the more abstract areas of experience by expressing them in concrete and familiar terms. In other words, metaphors give substance to conceptual knowledge by transferring knowledge from a known and concrete domain to an unknown and abstract one. The theory that Lakoff gradually built in partnership with philosopher Mark Johnson (Lakoff and Johnson 1980, 1987) postulates that metaphors surround us in our dailylifes not only through language, their main vehicle, but especially through our thoughts and actions. Let us take the sentence "our relationship has hit a dead-end street": in this sentence, love is seen as a journey and the relationship has reached a dead point, the two lovers cannot follow the street they wanted to explore and therefore have to choose whether it is better to go back or break the relationship in mutual agreement. This principle allows us to understand the conceptual domain of love by using the terms of a journey: the metaphor uses the understanding of a domain of experience (journey) to understand a more abstract domain (love). This is not a grammatical principle and it cannot be found in a vocabulary, but it is rather linked to the conceptual system underlying language.

From a technical perspective, metaphors can be defined as maps that connect a source domain to a target domain. This map is made of ontological correspondences that semantically connect entities in the source domain (in the case of love: lovers, their scopes, and struggles) to entities in the target domain (in the case of journey: travellers, vehicles, destinations, etc.). This map is therefore a collection of one-to-one relationships.

Maps are usually expressed with a propositional structure, for example "love is a journey", but they are not clauses. Confusing maps with their structures means falling into the trap of thinking that—according to this theory—metaphors are clauses: they are not, they are simple maps, collections of conceptual correspondences. Other examples of concepts directly shaped from experience and providing source domains to conceptual metaphors are the concepts of "object", "substance" and "container" (an example of the container metaphor is "the ship is coming into view") (Lakoff e Johnson 1980).

Lakoff and Johnson (1999) state that our conceptual system evokes a "container structure" on a variety of concepts that have nothing to do with containers. This happens because there are several concepts linked to experience—such as containers—that have such a clear and easily understandable structure that can be used to understand concepts that are less clear and intuitive. Sentences such as "to be in", "to live in a society", "to be there in 5 min", "to be in heat" are examples of metaphors linked to containers (time or emotional statuses are containers).

The most commonly used metaphors in the domain of emotions are related to emotions expressed as fluids in containers (Lakoff and Johnson 1980). In any container, the content cannot grow infinitely: there is a limit that once reached implies a change. In metaphors, the stronger the emotion, the bigger the volume of the container. There are metaphors linked to the idea of an elastic container that changes its physical status under pressure ("filled with pride", "filled with anger", etc.). The idea of getting filled suits emotions such as anger or pride, perhaps because the perceivable behaviour of the subject involved offers a strong experience. At the same time, this idea is also linked to bursts of anger and pride: even a flexible container has a limit and, when reached, it bursts open, often damaging itself.

There are other metaphors that relate to the possibility that containers are full and, therefore, overflow: "to be overwhelmed with love", "to shower somebody with love", "to be filled with joy". These expressions are often used to express affection, love, joy, which has indeed its specific version in the general metaphor "emotions as fluids in containers". In fact, joy is not expressed in terms of temperature (as is often the case for strong emotions), but rather in terms of movement of the fluid, using words such as "sparkling", "bubbling", which relate to something more dynamic (e.g. "to be always so happy and sparkling").

Other metaphors focus on the qualitative feature of what containers hold, stressing its purity. These metaphors relate to positive emotions and often contain adjectives that emphasize their positivity: "my feelings are pure". Joy is described using terms linked to a container by referring to the fact that the quantity of the liquid is very high and the container is full, overflowing, almost bursting. This fact leads humans to believe that joy is not common—our hearts cannot always be filled with joy—which makes the container metaphor an expression that highlights the uniqueness of this fact.

Thanks to Lakoff and Johnson's studies, metaphors become vehicles that make possible to provide a meaning and understand the world through language, giving humans a tool to organise their day-to-day conceptual reality.

5.3.1 Conceptual Metaphors and Mathematics

Among daily conceptual activities, there are also those related to quantity and—generally speaking—ideas such as "basic spatial relations, groupings, small quantities, motion, distributions of things in space, changes, bodily orientations, basic manipulations of objects (e.g., rotating and stretching), iterated actions, and so on." (Lakoff and Núñez 2000, p. 28).

For this reason, the book "Where Mathematics Comes From. How the Embodied Mind Brings Mathematics Into Being" (Lakoff and Núñez 2000) by Lakoff and Rafael Núñez, an expert in the teaching of mathematics, tried to apply the perspective given by conceptual metaphors to the domain of mathematics.

The authors contend that what makes mathematics a non-arbitrary discipline despite the fact of being created by humans—or more specifically, by their brains—is that it "uses the basic conceptual mechanisms of the embodied human mind as it has evolved in the real world" (Lakoff and Núñez 2000, p. 9).

The embodied paradigm—which lays the foundations of so-called second generation cognitive sciences—represents a radical change compared to first generation cognitive sciences, because it does not see the human mind as a mere processor of symbols (computationalism). By applying this new paradigm, the studies on cognition and learning moved from a perspective focussed on the abstract aspects of thought, which were governed by formal rules and were independent of cultural factors, to a perspective where the mind is rooted in the body, decentralised, action-oriented, holistic, culture-depended, deeply connected to biological principles. The guiding principle is that smart behaviour is an expression of biological bodies that act in their material and cultural environment, becoming drivers of its change (the key words of this new paradigm are "located", "diffused", "social", and "embodied", cfr. Hutchins 1996; Lave 1988; Varela et al. 1992).

According to the authors, by applying the embodied paradigm to mathematics, it is possible for the first time in history to account for the beauty and depth of mathematics, whereas previously the majority of the cognitive structures of mathematics were simply unknown. As they underline several times in their book:

> it is only with these recent advances in cognitive science that a deep and grounded mathematical idea analysis becomes possible. Insights of the sort we will be giving throughout this book were not even imaginable in the days of the old cognitive science of the disembodied mind, developed in the 1960s and early 1970s. In those days, thought was taken to be the manipulation of purely abstract symbols and all concepts were seen as literal-free of all biological constraints and of discoveries about the brain. Thought, then, was taken by many to be a form of symbolic logic (Lakoff and Núñez 2000, p. 5).

Lakoff and Núñez share the same starting assumptions of Dehaene in his *Number Sense*, i.e. the existence of an inborn-approximative form of mathematics in humans (and animals) that allow for subitizing numerosities under 4. Nevertheless, in Lakoff and Núñez, subitizing does not exhaust the arithmetical abilities of humans, who use counting to go beyond subitizing and achieve a certain degree of accuracy and reliability in their activities. Counting involves the following cognitive capacities:

(a) Grouping capacity.
(b) Ordering capacity.
(c) Pairing capacity.
(d) Memory capacity.
(e) Exhaustion-detection capacity.
(f) Cardinal-number assignment.
(g) Independent-order capacity.

Furthermore, the following skills are needed to count beyond 4:

(h) Combinatorial-grouping capacity.
(i) Symbolising capacity.

On the other hand, going beyond counting implies a conceptualisation ability that requires a much wider skill set and shares some traits with other conceptualisation tasks, such as the development of conceptual metaphors and the ability to make metaphorical blends. These skills allow to break the limits of subitizing and counting, leading to the understanding of arithmetic with natural numbers. It should nonetheless be noted that there are two types of conceptual metaphors that humans employ to venture in the domain of mathematical concepts: grounding metaphors and linking metaphors. The former are metaphors that allow humans to transpose daily experiences (e.g. putting the batteries in an object) into abstract concepts such as addition. The latter, instead, are metaphors that allow humans to link arithmetic tasks to other fields of mathematics. In the words of the authors (Lakoff and Núñez 2000, p. 53):

> (1) Grounding metaphors yield *basic, directly grounded ideas*. Examples: addition as adding objects to a collection, subtraction as taking objects away from a collection, sets as containers, members of a set as objects in a container. These usually require little instruction.

(2) Linking metaphors yield *sophisticated ideas,* sometimes called *abstract* Examples: numbers as points on a line, geometrical figures as algebraic equations, operations on classes as algebraic operations. These require a significant amount of explicit instruction.

One of the main ways in which conceptual metaphors play their pivotal role is by maintaining an image schema structure. Image schemas are topology-based and guiding structures that define spatial inferences. They are regular, dynamic, and repeat action patterns that order sequences, perceptions, and concepts. They derive from the physical experience of movement, object handling, and perceptive inter-actions (nota bene, perception should not be intended as the passive detection of experiences, but rather as an active process, including building, modelling, and ordering items). Among the image schemas available in mathematics, a major role is played by "container" schemas, which are based on the following inferential laws: the law of excluded middle, modus pones, modus tollens and hypothetical syllogism. Other important image schemas in mathematics are "source-path-goal", "above", "under", "contact", "support". The authors postulate that several basic mathematical concepts are built by combining these schemas.

By maintaining the image schema structure, grounding metaphors allow for the conceptualisation of numbers as sets, providing yet another logic structure to the broad concept of number. Grounding metaphors therefore enrich the notions of inborn arithmetic, keeping at the same time its properties. There are four grounding metaphors that are universal and do not depend on culture. These are:

(a) Arithmetic as object collection.

This metaphor goes from tangible objects (source domain) to numbers (target domain) and allows humans to perform additions and subtractions. As detailed by the authors, the metaphoric map is composed of the following elements (Lakoff and Núñez 2000, p. 55):

(a) the source domain of object collection (based on our commonest experiences with grouping objects),
(b) the target domain of arithmetic structured nonmetaphorically by subitizing and counting); and
(c) a mapping across the domains (based on our experience subitizing and counting objects in groups).

To perform multiplications and divisions, what is needed is a "metaphorical blend", a simultaneous activation of the metaphor and two domains (source and target), because these are complex tasks from a cognitive perspective. When humans perform these operations, they need to refer simultaneously both to numbers and object collections, for a repeated amount of times.

(b) Arithmetic as object construction.

This metaphor allows humans to understand fractions as parts of an object. As in the previous case, through the metaphorical blend, the scope of this metaphor can be broadened in two ways: merging/dividing and performing repeated additions and subtractions.

(c) Numbers are physical segments.

With this metaphor, natural numbers, the number zero and complex positive fractions (rational numbers) are defined in terms of tangible segments. As specified by the authors:

> The oldest land still often used) method for designing buildings or physically laying out dimensions on the ground is to use a measuring stick or string taken as a unit. These are physical versions of what in geometry arc called line segments. We will refer to them as "physical segments." A distance can be measured by placing physical segments of unit length end-to-end and counting them. In the simplest case, the physical segments are body parts: fingers, hands, forearms, arms, feet, and so on. When we put physical segments end-to-end, the result is another physical segment, which may be a real or envisioned tracing of a line in space. (Lakoff and Núñez 2000, p. 68)

Through the "metaphorical blend", the opposite procedure to the one previously described is possible (having a segment corresponding to a number), in order to create irrational numbers.

(d) Arithmetic as motion along a path.

This metaphor forces humans to imagine numbers as dynamic elements that go from one point to another on a straight line. The starting point of the motion is located on one extremity of the line; the destination point is on the other. This metaphor is used by the authors to introduce negative numbers, the points placed on the other side of the extremity where positive numbers are.

The four grounding metaphors are universal and they are naturally inferred from experience. They widen the domain of inborn arithmetic, which can be subitized, broadening the mathematical knowledge beyond natural numbers and venturing in the domains of other numbers. Nevertheless, what is still unclear is how the creation of these metaphors from random connections that get stabilised by repetition can lead to something so universal and globally shared. In other words, while the role played by metaphors in the creation of concepts may provide an interesting insight into how new concepts are created, it is still unclear what drives humans to create metaphors. A sound theory on concepts should meet the criteria highlighted by the authors (realism, evidence convergence, generality), but it seems that these refer to something that goes beyond the creation process of metaphors.

To conclude, also in this case (as Fodor flagged in Susan Carey's theory) there is a difficult "circularity" issue to address. Let us take the example of the "container" image schema. The authors postulate that this schema includes (in the sense that no deductive process is needed to come to a conclusion) logical expressions that can be inferred from the following statements (Lakoff and Núñez 2000, p. 31):

> 1. Given two Container schemas A and B and an object X, if A is *in* B and X is *in* A, then X is *in* B.

2. Given two Container schemas *A* and *B* and an object *Y*, if *A* is *in B* and *Y* is *outside of R*, then *Y* is *outside of A*.

These are self-evident statements that can be pictured as follows (Lakoff and Núñez 2000, p. 32):

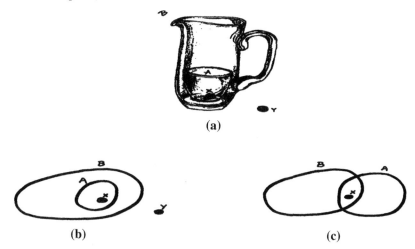

(a)

(b)   (c)

FIGURE 2.1   The logic of cognitive Container schemas. In (a), one cognitive Container schema, *A*, occurs inside another, *B*. By inspection, one can see that if *X* is in *A*, then *X* is in *B*. Similarly, if *Y* is outside *B*, then *Y* is outside *A*. We conceptualize physical containers in terms of cognitive containers, as shown in (b), which has the same logic as (a). However, conceptual containers, being part of the mind, can do what physical containers usually cannot—namely, form intersections, as in (c). In that case, an imagined entity *X* can be in two Container schemas *A* and *B* at once. Cognitive Container schemas are used not only in perception and imagination but also in conceptualization, as when we conceptualize bees as swarming *in* the garden. Container schemas are the cognitive structures that allow us to make sense of familiar Venn diagrams (see Figure 2.4).

The authors postulated that humans conceptualise physical containers as cognitive containers, i.e. subjects can see or use schemas through concrete objects. Cognitive containers become similar to Venn diagrams (cfr. description of Fig. 2.1, Lakoff and Núñez 2000), which are though concepts and mathematical techniques belonging to the 19th century, developed well before cognitive containers. For this reason, other authors (cfr. Lolli 2004) suspect that, in this case, the container schema is not used to explain a mathematical concept, but rather the other way around and, therefore, the container schema uses the mathematical concept of the end of the 19th century to explain how the actual concept of container works.

The doubts become even stronger considering that metaphors introduce objects in a target domain that, however, already exists and is well defined as a domain. For example, the authors contend that metaphors are useful to understand and perform tasks that are more difficult than the simple act of counting (they are used to introduce the concept of number). Nevertheless, counting requires the understanding of the concept of number (cardinal and ordinal alike). The same goes for the grounding metaphor "numbers as object collections", where addition and other actual tasks in the source domain require arithmetic abilities (e.g. the representation of a basis requires "the creation of subgroups". But these can be created only starting from a mathematical criterion).

The latter remark is highly interesting considering that Lakoff and Núñez established that metaphorical structures are important because they allow humans to understand the most obscure sides of mathematics thanks to the explanation of abstract concepts through the use of much more concrete concepts. Nevertheless, in the example mentioned above, the opposite seems to be true. It seems that concrete concepts (containers) can be understood only starting from much more abstract concepts (Venn diagrams). There are several other examples in mathematics and physics where the abstract concepts allow humans to understand and better clarify other concepts; just think of how much the concept of ideal gas helps in understanding the concept of real gas. In fact, the same happens in Lakoff and Núñez's book, when they repeatedly employ the concept of "metaphorical blends", a concept that is much more abstract than the one of grounding metaphor (it is not by chance that it is never specified whether they are "embodied" too, since they involve different domains and overlap at several levels). The same goes even for the grounding metaphor "arithmetic as object collection" and "classes as containers"; in both cases, the authors do not specify how the target domains of "integers" and "classes" are established.

Summing up, despite several studies proving over the years that the core of Lakoff and Núñez's theory is not a mere abstract construction, but is rather based on neurological and psychological evidence and, hence, that metaphors truly play a paramount role in some understanding processes (cfr. Gibbs 2005; Dodge and Lakoff 2005; Gallese and Lakoff 2005), the idea of "conceptual metaphor" was strongly criticised, focussing in particular on the conceptual correspondence between different domains and the processing of the elements in these domains (as well as their level of specificity). According to several authors (cfr. Casadei 1999, 2003; Glucksberg and Keysar 1993; Ortony 1993; Sperber and Wilson 2006; Wilson and Carston 2006), the main limitation of conceptual metaphors is the so-called "conceptual reductionism" put forward by Lakoff, consisting in ignoring the role of language in the conceptualisation processes related to the processing and structuring of the "process of thought", while focussing only on the issues related to mental metaphors.

The different critics of the theory share the concern that understanding what conceptualising through metaphors means is important, because if conceptualising means "creating concepts", then assessing the theoretical reasoning related to this knowledge-building mechanism is also important. It goes indeed without saying that humans have only one tool at their disposal to transcend the limits of their knowledge and understand what it is still unknown to them: use what they know. From this perspective, metaphors are important because they describe the reality we are about to discover by using what we already know. Nevertheless, it is logical that metaphors do not completely deplete the toolbox at our disposal when we wish to learn something new. For example, equally efficient techniques are induction, hybridisation, analogy, metonymy, abduction, etc. This is especially true for mathematical concepts (cfr. Cellucci 2013), where the content available up to a certain moment and ready to be used and applied to what is still unknown can easily have nothing to do with metaphors.

Lakoff and Núñez couple conceptual metaphors (universal) with other knowledge-building mechanisms (cultural), which—once combined—allow for the modelling of some aspects of reality. The authors' aim is to recognise the potential of language and culture to provide non-universal metaphors that provide inputs to explain the stability of mathematics. The use of these cultural metaphors explains the non-monolithic structure and organisation of mathematics in its disciplines, which can keep an internal coherence despite being opposed one to the other. Cultural metaphors of this kind are (Lakoff and Núñez 2000, p. 358):

1. The essence of a subject matter is to be given by a small set of axioms.
2. Mathematical reasoning is a form of mathematical calculation, which allows all mathematical truths to be calculated using mathematical logic).
3. All respectable subject matters can and must have secure foundations on which everything in the subject matter is built.

The mistake lies in interpreting literally what is a simple cultural idea created outside the mathematical domain and inside a specific cultural and historical framework. According to the authors, this was the mistake made by all three schools belonging to the so called "Movement for the Foundations of Mathematics", which uselessly strived to infer all mathematical truths from the axioms of logics. The disastrous consequence of all this was that:

> The Foundations movement itself collapsed. None of these ideas has stood the test of time within mathematics. All three have been found to be mathematically untenable. But they all shaped the structure of mathematics itself not just how it was done but its very content. One cannot even imagine contemporary mathematics without these ideas. (Lakoff and Núñez 2000, p. 358)

The foundational schools failed in their task because their ideas had nothing to do with the structure of the universe. Their concepts were not "built into the brain structure of all human beings, as basic numeration is. And they are not cognitive universals. They are, rather, products of human culture and human history" (Lakoff and Núñez 2000, p. 358).

An embodied mathematics based on conceptual metaphors is therefore the only available solution to avoid the mistakes made by the foundationalists. It is the only theory that avoids these mistakes because it is built on the cognitive mechanisms that are at the basis of our knowledge. Nevertheless, as outlined in this paragraph, conceptual metaphors cannot solve all problems and, above all, do not explain how the cognitivist approach can overcome the limits that the two authors identified in the foundationalist approach. This is the topic that we will address in the following paragraph.

5.4 Intuitionism and Cognitive Sciences

As previously stated, Lakoff and Núñez do not directly address the philosophical question related to foundationalist schools, but they simply challenge their "absolute" character, which aimed at inferring all mathematical truths from logic. Delving into the details, the authors obviously recognise that their endeavour aims at finding an answer to the question of the origins of mathematics, which has always been at the centre of the interest and discussion of philosophers.

Nevertheless, their answer tries to reject the ideas of the "philosophical environment" theory, by postulating that its "Romantic Conception of Mathematics" should be replaced by embodied mathematics. Therefore, following Lakoff and Núñez's ideas to the letter, the approaches highlighted by foundationalist schools during the previous century should be considered as completely inappropriate and obsolete. Nevertheless, this is not the case. In fact, all approaches (formalism, logicism, and intuitionism) are an inescapable point of reference when framing the new studies on the philosophy of mathematics. Several authors contended that the main issue of the philosophy of mathematics was and still is the "foundation of mathematics", meaning its "metaphysical, epistemic, and mathematical" foundations (Shapiro 2004, p. 37). This may be the reason why some cognitive scientists —and not only those more philosophy-oriented (e.g. Lakoff and Núñez)—joined the debate on foundationalist approaches. The same goes for Stanislas Dehaene, who directly challenged the foundational schools by taking the side of intuitionism. In his words:

> Among the available theories on the nature of mathematics, intuitionism seems to me to provide the best account of the relations between arithmetic and the human brain. The discoveries of the last few years in the psychology of arithmetic have brought new arguments to support the intuitionist view that neither Kant nor Poincaré could have known. These empirical results tend to confirm Poincaré's postulate that number belongs to the "natural objects of thought," the innate categories according to which we apprehend the world." (Dehaene 2011, pp. 226–27).

Taking into account the studies that confirmed that children are born with mechanisms allowing them to identify objects and extract numerosities from small sets, and that animals as well have this "inborn" and language-independent number sense, the French neuroscientists defines his research activity as a "Kantian research agenda" that aims at understanding how the intuitions that are at the basis of experience are possible, on which neuronal structures they are based, and how these intuitions can be modified through education and learning (Dehaene and Brannon 2011). In the words of the author:

> from grid cells to number neurons, the richness and variety of the mechanisms used by animals and humans, including infants, to represent the dimensions of space, time and number is bewildering and suggests evolutionary processes and neural mechanisms which may universally give rise to Kantian intuitions (Dehaene and Brannon 2011, p. iX).

Following a huge number of studies, Dehaene contends to have enough arguments to launch an ambitious reformulation of all the questions that traditionally were tackled in the field of the philosophy of mathematics, taking into account the current debate in cognitive sciences and starting from a sound Kantian philosophical basis (Graziano 2015).

Nevertheless, by doing so, he ends up making patent mistakes, such as taking Kant's "a priori knowledge" as "inborn knowledge", which is present in subjects since their birth and represents a "potential" ready to be developed on the condition that suitable environmental conditions exist.

From this perspective, "inborn" is used as synonym for "immutable", which leads Dehaene in misunderstanding it for "a priori". Differently to Kant's "a priori" knowledge, Dehaene's inborn knowledge is not experience-independent. In fact, it can be acquired by formulating hypotheses based on assumptions that can be inferred from experience and the plausibility of these assumptions and conclusions can be evaluated by comparing them to experience. Furthermore, this knowledge is not immutable; future exceptions are possible, maybe due to very slow changes governed by evolutionary laws. They are not even intrinsically necessary, but rather contingent, since they could be (in)compatible with future data. Finally, they are not certain, since there is no guarantee that in future there will not be counterexamples.

In addition to this, saying that knowledge is "a priori" in Kantian sense means providing a logic statement and not a description. In other words, it does not mean that knowledge is independently perceivable from a specific experience. It is not a statement related to the functioning of reality. On the opposite, it suggests that knowledge logically precedes the possibility that every experience is as it is. In Kant, "a priori" is synonym for "transcendental" and deals with the possibility of an experience, not its more or less detailed description.

Dehaene's view on inborn knowledge differs both from Kant's ideas and the character of absolute perfection of mathematics as an absolutely certain a priori knowledge. In a Kantian perspective, intuition is the only source of certain knowledge-based mathematics, because this knowledge has not been processed through conceptual mediation, it has not become a "judgment". On the contrary, Dehaene's intuitive mathematics is not infallible, because it is not an absolutely certain a priori knowledge. In fact, it has a narrower connotation, far from infallible, and subject to the psychological laws such as the distance and size effects.

These considerations should clarify that Dehaene's positions—which he considers part of a "Kantian research agenda"—are quite different from Kant's philosophical ideas. Furthermore, they differ also from the ideas of another author he often quotes in *The Number Sense*, i.e. Poincaré. Also Poincaré uses "intuition" differently from Dehaene (Graziano 2013). As he writes:

> We have then many kinds of intuition; first, the appeal to the senses and the imagination; next, generalization by induction, copied, so to speak, from the procedures of the experimental sciences; finally, we have the intuition of pure number, whence arose the second of the axioms just enunciated, which is able to create the real mathematical reasoning. (Poincaré 1907, p. 20)

According to Poincaré, only the intuition of pure numbers is certain, because they are the statement of a property of intelligence. Therefore, when Dehaene states that he shares Poincaré's intuitionism (comparing it to his early notion of subitizing), he does not consider that Poincaré uses it as a demonstration tool and not only as a number generator. As a matter of fact, despite starting from an intuitionist perspective, Poincaré philosophical stance includes without any doubts some formalisation elements.

Taking a look to Poincaré's famous passage "Thus logic and intuition have each their necessary role. Each is indispensable. Logic, which alone can give certainty, is the instrument of demonstration; intuition is the instrument of invention" (Poincaré 1907, p. 23), it is possible to notice that it only summarises a very important argument of his essay *The value of science*, in which he dwells upon the dialectic existing between sensitive intuition and analytical procedures, which Poincaré calls verifications and which are based on syllogism, replacement and nominal definition. It was precisely because of these remarks that the positions of Poincaré were defined as belonging to a kind of semi- or pre-intuitionism.

Because of the remarks on Poincaré's concept of intuition, some authors put forward the idea that Dehaene rather shares the starting assumption formulated by the founder of intuitionist mathematics, i.e. Brouwer (Longo 2005). Dehaene seems indeed to share the explicitly non-linguistic character of Brouwer's mathematics, in other words the belief that language does not play a role in building mathematical concepts, because mathematics is based on a free and creative activity of humans founded on the intuition of time. Brouwer contends that mathematics is a mental construction based on the "Primordial Intuition", which makes individuals aware of the different components of time, i.e. two discrete entities, one present and one past.

In other words, when PI occurs, the consciousness retains two neighbouring elements that are different but unified. The two elements are not identical, but they do form a single unit that Brouwer calls "twoity". It is worth noting that when dealing with PI, individuals are affected by the action of consciousness, which is described by Brouwer in different ways, not so much as an event but rather as a phenomenon or a process. Primordial intuition can be endlessly repeated; it depends only on the free will of the subject, thereby producing sequences of increasingly complex mental objects simply as repetition of the primordial act. With a twoity, it is possible to build a threeity; with a threeity it is possible to build another construction, and so on. Therefore, according to Brouwer, all numbers—ordinal, natural and other—are constructions obtained from reiterations of PI: "This intuition of two-oneness, this ur-intuition of mathematics, creates not only the numbers one and two, but also all finite ordinal numbers" (Brouwer 1912, p. 12).

Therefore, according to Brouwer, the mental construction made by individuals starting from the Primordial Intuition precedes the linguistic description.

This separation of the language of mathematics from mathematics is the subject of the First Act Of Intuitionism:

> Completely separating mathematics from mathematical language and hence from the phenomena of language described by theoretical logic, recognizing that intuitionistic mathematics is an essentially languageless activity of the mind having its origin in the perception of a move of time. This perception of a move of time may be described as the falling apart of a life moment into two distinct things, one of which gives way to the other,

but is retained by memory. If the twoity thus born is divested of all quality, it passes into the empty form of the common substratum of all twoities. And it is this common substratum, this empty form, which is the basic intuition of mathematics. (Brouwer 1981, pp. 4–5)

Without going into the details of the role of language, it seems indeed true that Dehaene provides neurophysiological evidence to Brouwer's ideas, because he gives experimental support to the fact that natural numbers are rooted in our brains and mathematics can be performed without language.

Nevertheless, according to Dehaene, alinguisticity is at the basis of elementary and approximative mathematics, which humans share with other animals. So, despite both authors sharing the need for a non-language based mathematics, for Brouwer it is something that deals with the whole mathematics, while for Dehaene it deals only with approximative mathematics, hence accessible also to animals, newborns, and all the people who do not have a wide vocabulary for numbers. According to Brouwer, the concept of mathematics is much broader than Dehaene's, because it includes all mental construction processes, conscious and unconscious, since Brouwers believes that all mental abilities are PI-based. In these terms, it is easy to decouple Brouwer's concept of pure mathematics—which corresponds to Dehaene's formal mathematics—from general mathematical abilities, which are the equivalent of our general cognitive skills.

In Dehaene's opinion, exact mathematics is symbolism, hence language, i.e. everything that in Brouwer's opinion should be avoided because it is strictly linked to outward-looking activities. Contrary to Dehaene, Brouwer believes that language does not ensure exact mathematics: according to him, mathematics, its truth and accuracy are to be found in the mental act.

The two authors also have a different opinion on the role of space and body. Dehaene strongly focusses on the bodily component of mathematical knowledge, while Brouwer contended that the body only plays an ancillary role (he even rejected its role in principle, starting from his mystical definition of pursue of happiness). Dehaene supported the fundamental role played by the body because he strongly trusted the data obtained in the study on non-Western populations that had a very limited mathematical vocabulary (cfr. Chapter 2). It is therefore clear that an intuitionist stance *à la Brouwer* strongly clashes with the second-generation embodied cognitive sciences heralded by Dehaene. Generally speaking, it is easy to say that Brouwer and Dehaene do not only have similarities, but rather strong differences.

To conclude, it is possible to say that the repeated "philosophical blunders" of Dehaene are due to the use of the term "intuition", which plays an important role in mathematics but is still a highly polysemic term that acquires different meanings according to the context in which it is used (as it also happens with the term "metaphorical"). For example, a clear distinction can be drawn between the concept of intuition in mathematical praxis and the same concept in the building and development of mathematics.

Among the different traditional meanings given to "intuition", there is without any doubt also the one related to an immediate, direct, and inferential mediation-free form of knowledge. This is the meaning that most attracts Dehaene, which pushes him to say that even mathematicians in the first stages of their work have claimed to possess a direct perception of mathematical relations. They say that in their most creative moments, which some describe as "illuminations", they do not reason voluntarily, nor think in words, nor perform long formal calculations" (Dehaene 2011, p. 136).

Nevertheless, it is well known that this approach to intuition had to face the pitiless criticism of Dieudonné, who stated that:

> the intuition of the whole is a great mystification, because no one I know has insight in the true sense of intuition, that is, immediate knowledge of whole numbers greater than ten. Consequently, to say that you have intuition of integers greater than ten is a big fraud" (Dieudonné 1981, p. 23).

Dehaene seems to escape Dieudonné's criticism because, while accepting the idea that intuition is immediate, direct, non-linguistic knowledge, he compares this concept of intuition to his idea of subitizing which, as we have seen, is only valid for the first three positive integers and that after three proves to be fallible and subject to the distance effect and size effect.

Some academics of the philosophy of mathematics share Dehaene's position and contend that intuition is basically fallible, even if it is at the basis of all reasoning and therefore necessary from a cognitive point of view. Robert Hanna is an example:

> Ninth, intuition is fallible, which is to say that it is always possible for an intuition to be wrong. Neither the authoritativeness of intuition nor its cognitive indispensability implies that it cannot be mistaken. Unfortunately for creatures with minds like ours, it is built into the cognitivist existential predicament (see Sect. 6.4) that the world might be otherwise than I take it to be, no matter how intrinsically compelling the evidence for my belief is. It is plausible to hold, given the authoritativeness of intuition together with its cognitive indispensability, that an intuition that S provides reliable evidence for the intuiting subject's belief that necessarily S. But even assuming this, an intuition that S cannot provide an epistemic guarantee that necessarily S. (Hanna 2006, p. 172)

Nevertheless, admitting the fallibility of intuition does not solve all problems; it rather creates an additional issue compared to an infallible and immediate intuition. If intuition is fallible, how can one know if intuition S is right or wrong? Intuition does not provide an answer, because it is fallible, and hence only reasoning can help. Nevertheless, if reasoning starts from a fallible form of intuition, how can one know if this principle is right or wrong? Once again, a circularity issue arises and an infinite loop is created.

References

Amalric, M., & Dehaene, S. (2016). Origins of the brain networks for advanced mathematics in expert mathematicians. *Proceedings of the National Academy of Sciences of the United States of America, 113*, 4909–4917.

Atkinson, J., Campbell, F. W., & Francis, M. R. (1976). The magic number 4 ± 0: A new look at visual numerosity judgments. *Perception, 5,* 327–334.

Baroody, A. J. (1984). More precisely defining and measuring the order-irrelevance principle. *Journal of Experimental Child Psychology, 38,* 33–41.

Baroody, A. J., & Gannon, K. E. (1984). The development of the commutativity principle and economical addition strategies. *Cognition and Instruction, 1,* 321–339.

Beck, J. (2017). Can bootstrapping explain concept learning? *Cognition, 158,* 110–121.

Briars, D., & Siegler, R. S. (1984). A feature analysis of preschoolers' counting knowledge. *Developmental Psychology, 20,* 607–618.

Brouwer, L. E. J. (1912). Intuitionism and formalism. In P. Benacerraf & H. Putnam (Eds.), *Philosophy of mathematics. selected readings,* (1964). Cambridge: Cambridge University Press.

Brouwer, L. E. J. (1981). *Brouwer's Cambridge lectures in intuitionism.* Cambridge: Cambridge University Press.

Bruner, J. S. (1968). *Processes of cognitive growth: Infancy.* Worcester, MA: Clark University Press.

Butterworth, B. (1999). *The mathematical brain.* London: Macmillan.

Carey, S. (2001). Cognitive foundations of arithmetic: Evolution and ontogenesis. *Mind and Language, 16*(1), 37–55.

Carey, S. (2004). Bootstrapping and the origin of concepts. *Daedalus, 133,* 59–68.

Carey, S. (2009a). *The origin of concepts.* New York: Oxford University Press.

Carey, S. (2009b). Where our number concepts come from. *The Journal of Philosophy, 106,* 220–254.

Casadei, F. (1999). Alcuni pregi e limiti della teoria cognitivista della metafora. *Lingua e Stile, XXXIV*(2), 167–180.

Casadei, F. (2003). Per un bilancio della Semantica Cognitiva. In L. Gaeta & S. Luraghi (Eds.), *Introduzione alla Linguistica Cognitiva* (pp. 37–55). Roma: Carocci.

Cellucci, C. (2013). *Rethinking logic: Logic in relation to mathematics, evolution, and method.* Dordrecht, Heidelberg, New York, London: Springer.

Chomsky, N. (1957). *Syntactic structures.* The Hague, Paris: Mouton.

Cooper, D. E. (1986). *Metaphor.* Oxford: Blackwell.

Dehaene, S. (1997). *The number sense. How the mind creates mathematics.* New York: Oxford University Press.

Dehaene, S. (2011). *The number sense. How the mind creates mathematics.* Revised and Updated Edition. New York: Oxford University Press.

Dehaene, S., & Brannon, E. (2011). *Space, time and number in the brain. Searching for the foundations of mathematical thought.* Amsterdam: Elsevier Academic Press.

Dieudonné, J. (1981). *History of functional analysis.* Amsterdam: North-Holland.

Dodge, E., & Lakoff, G. (2005). Image schemas: From linguistic analysis to neural grounding. In B. Hampe (Ed.), *From perception to meaning: Image schemata in cognitive linguistics* (pp. 57–91). Berlin-New York: Mouton de Gruyter.

Feigenson, L., Dehaene, S., & Spelke, E. S. (2004). Core systems of number. *Trends in Cognitive Sciences, 8,* 307–314.

Feigenson, L., Carey, S., & Spelke, E. (2002). Infants' discrimination of number vs. continuous extent. *Cognitive Psychology, 44,* 33–66.

Fillmore, C. (1985). Frames and the semantics of understanding. *Quaderni di Semantica, 6,* 222–254.

Fodor, J. A. (2008). *The language of thought revisited.* New York: Oxford University Press.

Fuson, K. C. (1988). *Children's counting and concepts of number.* Berlin: Springer-Verlag.

Fuson, K. C., & Hall, J. W. (1983). The acquisition of eraly number word meanings. In H. Ginsburg (Ed.), *The development of children's mathematical thinking* (pp. 49–107). New-York: Academic Press.

Fuson, K. C., Pergament, G. G., & Lyons, B. G. (1985). Collection terms and preschoolers' use of the cardinality rule. *Cognitive Psychology, 17,* 315–323.

Gallese, V., & Lakoff, G. (2005). The brain's concepts: The role of the sensory-motor system in conceptual knowledge. *Cognitive Neuropsychology, 22,* 455–479.

Gallistel, C. R., & Gelman, R. (1992). Preverbal and verbal counting and computation. *Cognition, 44,* 43–74.

Geary, D. C. (1994). *Children's mathematical development: Research and practical applications.* Washington, DC: APA.

Gelman, R. (1993). A rational-constructivist account of early learning about numbers and objects. In D. Medin (Ed.), *The psychology of learning and motivation.* San Diego, CA: Academic Press.

Gelman, R., & Gallistel, C. R. (1978). *The child's understanding of number.* Cambridge, MA: Harvard University Press.

Gelman, R., Greeno, J. G. (1989). On the nature of competence: Principles for understanding in a domain. In L. B. Resnick (Ed.), *Knowing and learning: Issues for a cognitive science of instruction,* 125–186. Hillsdale, NJ: LEA.

Gelman, R., & Meck, E. (1983). Preschooler's counting: Principles before skills. *Cognition, 13,* 343–359.

Gelman, R., & Meck, E. (1986). The notion of principle: The case of counting. In J. Hiebert (Ed.), *Conceptual and procedural knowledge: The case of mathematics* (pp. 29–57). Hillsdale, NJ, US: Lawrence Erlbaum Associates.

Gelman, R., Meck, E., & Merkin, S. (1986). Young children's numerical competence. *Cognitive Development, 1,* 1–29.

Gelman, R., & Tucker, M. F. (1975). Further investigations of the young child's conception of number. *Child Development, 46,* 167–175.

Gibbs, R. W. (2005). The psychological status of image schemas. In B. Hampe (Ed.), *From perception to meaning: Image schemata in cognitive linguistics* (pp. 113–135). Berlin-New York: Mouton de Gruyter.

Glucksberg, S., & Keysar, B. (1993). How metaphor works. In A. Ortony (Ed.), *Metaphor and thought* (pp. 401–424). Cambridge: Cambridge University Press.

Graziano, M. (2013). Numbers in mind. Between intuitionism and cognitive science. *Reti, Saperi, Linguaggi, 4,* 72–79.

Graziano, M. (2015). Numerical cognition and philosophy of mathematics. Dehaene's (neuro) intuitionism and the relevance of language. *Rivista Italiana di Filosofia del Linguaggio,* SFL 2014, 362–377.

Groen, G. J., & Parkman, J. M. (1972). A chronometric analysis of simple addition. *Psychological Review, 79,* 329–343.

Halford, G. S. (1993). *Children's understanding: The development of mental models.* Hillsdale, NJ: Erlbaum.

Hanna, R. (2006). *Rationality and logic.* Cambridge, MA: The MIT Press.

Hespos, S. J., & Spelke, E. S. (2004). Conceptual precursors to language. *Nature, 430*(6998), 453–456.

Hutchins, E. (1996). *Cognition in the wild.* Cambridge: MIT Press.

Lakoff, G., & Johnson, M. (1999). *Philosophy in the flesh.* New York: Perseus Book.

Lakoff, G., & Johnson, M. (1980). *Metaphors We Live By.* Chicago: Chicago University Press.

Lakoff, G., & Johnson, M. (1987). *Women, fire and dangerous things: What categories reveal about the mind.* Chicago: University of Chicago Press.

Lakoff, G. (1987). *Women, fire, and dangerous things.* Chicago: University of Chicago Press.

Lakoff, G., & Núñez, R. E. (2000). *Where mathematics comes from: How the embodied mind brings mathematics into being.* New York: Basic books.

Langacker, R. W. (1987). *Fondations of cognitive grammar, vol 1: Theoretical prerequisites.* Stanford: Stanford University Press.

Lave, J. (1988). *Cognition in practice. Mind mathematics and culture in everyday life.* Cambridge: Cambridge University Press.

Lépine, R., Barrouillet, P., & Camos, V. (2003). *New evidence about the nature of the subitizing process*. The XIII Conference of the European Society for Cognitive Psychology. Grenade, Espagne.

Lolli, G. (2004). *Da dove viene la matematica. Recensione di G. Lakoff, R. E. Nunez, Where Mathematics comes from*. http://homepage.sns.it/lolli/articoli/LakNun.pdf.

Longo, G. (2005). The cognitive foundations of mathematics: Human gestures in proofs and mathematical incompleteness of formalisms. In M. Okada, et al. (Eds.), *Images and reasoning* (pp. 105–134). Tokio: Keio University Press.

Majid, A., Bowerman, M., Kita, S., Haun, D. B., & Levinson, S. C. (2004). Can language restructure cognition? The case for space. *Trends in Cognitive Science, 8*(3), 108–114.

Mandler, G., & Shebo, B. J. (1982). Subitizing: An analysis of its component processes. *Journal of Experimental Psychology: General, 111*, 1–22.

Molko, N., Wilson, A., & Dehaene, S. (2004). Dyscalculie, le sens perdu des nombres. *La Recherche, 10*, 42–49.

Ortony, A. (1993). *Metaphor and thought*. Cambridge: Cambridge University Press.

Piazza, M., Fumarola, A., Chinello, A., & Melcher, D. (2011). Subitizing reflects visuo-spatial object individuation capacity. *Cognition, 121*(1), 147–153.

Poincaré, H. (1907). *The value of science*. New York: The Science Press.

Pylyshyn, Z. W. (1998). The role of visual indexes in spatial vision and imagery. In R. Wright (Ed.), *Visual attention* (pp. 215–231). Oxford: Oxford University Press.

Resnick, L. B. (1986). The development of mathematical intuition. In M. Perimutter (Ed.), *Perspectives on intellectual: The Minnesota symposia on child psychology*. Hillsdale, NJ: LEA.

Revkin, S. K., Piazza, M., Izard, V., Cohen, L., & Dehaene, S. (2008). Does subitizing reflect numerical estimation? *Psychological Science, 19*(6), 607–614.

Rey, G. (2014). Innate and learned: Carey, mad dog nativism, and the poverty of stimuli and analogies (yet again). *Mind and Language, 29*, 109–132.

Rips, L. J., Bloomfield, A., & Asmuth, J. (2008). From numerical concepts to concepts of number. *Behavioral and Brain Sciences, 31*, 623–642.

Rosch, E. (1975). Cognitive representations of semantic categories. *Journal of Experimental Psycology, 104*, 192–233.

Searle, J. R. (1979). *The metaphor*. Cambridge: Cambridge University Press.

Shapiro, S. (2004). Foundations of mathematics: Metaphysics, epistemology, structure. *The Philosophical Quarterly, 54*, 16–37.

Shipley, E. F., & Shepperson, B. (1990). The what-if of counting. *Cognition, 36*, 285–289.

Siegler, R. S., & Crowley, K. (1994). Constraints on learning in nonprivileged domains. *Cognitive Psychology, 27*, 194–226.

Sperber, D., & Wilson, D. (1986). *Relevance: Communication and cognition*. Oxford: Blackwell.

Sperber, D., & Wilson, D. (2006). A deflationary account of metaphor. UCL Work. *Pap. Linguist, 18*, 171–203.

Svenson, O. (1975). Analysis of time required by children for simple addition. *Acta Psychologica, 39*, 289–302.

Trick, L. M., & Pylyshyn, Z. W. (1993). What enumeration studies can show us about spatial attention: Evidence for limited capacity preat-tentive processing. *Journal of Experimental Psychology: Human Per- ception and Performance, 19*, 331–351.

Varela, F. J., Thompson, E. T., & Rosch, E. (1992). *The embodied mind: Cognitive science and human experience*. Cambridge, MA: The MIT Press.

Xu, F. (2003). Numerosity discrimination in infants: Evidence for two systems of representations. *Cognition, 89*, 15–25.

Wilson, D., & Carston, R. (2006). Metaphor, relevance and the 'Emergent Property' issue. *Mind and Language, 21*(3), 230–260.

Wittgenstein, L. (1953). *Philosophical investigations*. Oxford: Blackwell.

Wynn, K. (1990). Children's understanding of counting. *Cognition, 36*, 155–193.

Chapter 6
Dual Process Theories for Calculus

Abstract The dual process theories are popular in many domains of psychology, such as reasoning, decision making, social cognition, cognitive development, clinical psychology, and cognitive neuroscience. In the last chapter, this theoretical approach is applied, for the first time, to the studies on numerical cognition with the aim of review the results brought about by psychological and neuroscientific studies conducted on numerical cognition and laying the foundations of a new potential philosophical explanation on mathematical knowledge.

Keywords Dual process theories · Maddy scientific naturalism
Liberal naturalism · Evolution

6.1 Biological Evolution and Cultural Evolution ·

The previous chapter outlined the reasons why it is fair to say that Dehaene and the other cognitive scientists that tried to find a form of continuity between the natural approximative system (system 1) and the exact cognitive system (system 2) failed in their endeavour. Despite the fact that the discovery of an inborn, biologically based, and animal-human shared form of mathematics might seem a convincing argument from a neuroscientific point of view—even more so by taking into consideration the cognitive architecture at its basis—this concept unfortunately exclusively applies to the concept of "numerosity", not to the concept of "number".

The term numerosity refers to the mere perceptive evaluation of different sets of items and the reasonable ability to compare them with bigger or smaller sets; in other words it refers simply to non-symbolic cardinality. On the other hand, in mathematics, numbers represent abstract entities with specific features (e.g. completeness for real numbers). Furthermore, they are represented by specific oral or written symbols and they can be used for computation. According to their features, they can be classified as natural, real, imaginary, negative, whole, rational, irrational, complex, etc. While growing up, humans learn how to use numbers through symbols in their various forms and in relation to other cognitive skills.

© The Author(s) 2018
M. Graziano, *Dual-Process Theories of Numerical Cognition*,
SpringerBriefs in Philosophy, https://doi.org/10.1007/978-3-319-96797-4_6

The core of the issue is that in order to learn the concept of number, it is necessary to have a word or symbol that refers to this concept. As it is widely known, in philosophy, Aristotle was the first to present this thesis by suggesting that we perceive numbers through the denial of the continuous [Ό δ αριθμος < αισθανομεθα > τη αποφασει του συνεχους] (DA, 425a 19). Franco Lo Piparo, an Italian philosopher of language, rightly pointed out that the translation of the term αποφασις with "denial" tempered the reference made by Aristotle to the language-based nature of the negation of continuous. In the words of the author:

> It should not be seen as the 'elimination' of something but as a linguistic act that states that something is different: ἀπόφασις comes from ἀπόφημι, which means 'I say no', 'I say that this is not that'. A number, because of its 'multiplicity of units', represents discontinuous entities perceivable following a cognitive operation that, describing the units used to identify them, differentiate them. A number therefore requires a linguistic negation act. (Lo Piparo 2014, p. 186)

To clarify even more this concept, Lo Piparo presents also the difference made by Aristotle between "unit" and "point":

> To highlight even more the linguistic ontology of a number, Aristotle stresses the importance of avoiding confusion between *unit* and *point*. While a *unit* is a result of a linguistic operation ("the negation of continuity"), a *point* is a physical sign of a concrete act. "A point and every division and whatever is indivisible in this way is made clear like a *privation*" (*De An.*, 430b 20–21). In other words: a *point* is the product of a subtraction from a physically wellDefault—Interventionistdelimited entity. This does not necessarily require a linguistic intervention. A point exists in the frame of a physical space, while a unit exists merely at a cognitive-linguistic level. (Lo Piparo 2014, p. 187)

Therefore, following Lo Piparo's reasoning (and his reference to Aristotle), units are the linguistic operations needed to develop quantification abilities, which lay at the basis of all other arithmetical concepts. It is hence through language that humans learn how to label an infinite amount of numbers, use symbols and distinguish between quantities. Consequently, the structure of mathematics becomes more abstract and refined. The establishment of an oral and written symbolic system, the ability to label an infinite amount of different numbers and to process continuous quantities as discrete, coupled with the possibility of inventing rules useful for arithmetic computation, are all products of cultural evolution.

Cultural evolution is subject to a much faster development pace than biological evolution, because it does not require cumulative genetic mutations selected through the passage from one generation to the other. It exploits the individual learning abilities allowed by the brains structure and the ability to convey knowledge to others by using symbolic language.

Nevertheless, contrary to the cultural evolution of mathematics, the biological evolution of our brains virtually halted after the emergence of the *Homo Sapiens*, approximately 100,000 years ago. As a result, cultural objects such as words and numbers are nowadays processed by biological systems originally designed for other tasks. This is the reason why humans are so inclined to approximation and struggle so much in learning the multiplication table by heart or computation with fractions. It is as if our brains refused to yield to these objects that go "against nature". This is also the reason why the mathematical objects that have a structure that suits our brains architecture seem so intuitive and easily recognisable (Dehaene 2011).

According to this perspective, humans are genetically much more similar to their closest relatives, primates, and therefore their sharing a good number of the basic cognitive tools at their disposal does not come as a surprise. Nevertheless, besides these common tools, humans own a wide network of complex cognitive skills that are unique and include symbolic communication, advanced reasoning abilities, or the ability to develop technologic tools and use them.

However, it would be a mistake to oppose cultural evolution and biological evolution, since the former is an enhancement and development of the latter. This is due to the fact that for the majority of their evolutionary history, humans have had to face problems similar to those of other creatures and they have had to struggle to survive and adapt to the surrounding environment as much all the other creatures. Later, thanks to cultural evolution, the situation changed, allowing individuals to invest less energy into the mere task of surviving. Nevertheless, in order to survive, humans still had to control their environment and try to enhance it (Cellucci 2013).

According to this hypothesis that postulates the existence of a continuum between cultural and biological evolution, in order to interpret cultural phenomena it is necessary to refer to the bio-cognitive conditions at their basis. Following this reasoning, interpretation models must be developed taking into account the knowledge coming from different disciplines, hence creating a continuum of knowledge, with biology on top of the list of the contributing disciplines. The cultural nature that characterises humans has to be seen as the product of our natural evolutionary history. In a nutshell: humans and human knowledge are just one (rather big) part of nature.

It is worth highlighting that supporting the continuum hypothesis does not mean saying that cultural evolution is a form of biological evolution. Despite the fact that biological evolution shaped organisms in order to overcome similar struggles to those that they had to face and overcome in an ancient past, the existing environment forces individuals to face new challenges that they cannot always overcome by using the tools provided by biological evolution, therefore exploiting the powerful tools that only cultural evolution can offer. The continuum hypothesis that postulates bio-cognitive constraints to cultural knowledge does not affect the content of knowledge. In the case of mathematics, humans have developed mathematical skills that are not simply fit for the purposes of specific tasks (as it is the case of non-human animals), but also more generic tasks.

Cultural and biological evolution are compatible, but do not overlap and, therefore, in order to allow for an exchange between these two entities, it is necessary to find a common ground, an intermediate level, which is provided by the "mind". Introducing the concept of mind in our discussion implies claiming that the creation, transmission, and evolution of knowledge has to be explained through the study of mental processes. We can understand the features characterising the *Homo Sapiens* as a species—as well as the structural and functional differences that it presents compared to other animal species—by describing the nature of our mind, as well as exploring the mechanisms underlying the cognitive processes that allow us to perceive measurements and spatial, temporal, and numerical sizes, providing us with the knowledge about the world that we need in order to act.

Nevertheless, we should be cautious as not to mix up the concept of mind and the processes taking place in the "dark" of our heads; in other words our brains' information processing architecture. Talking about the mind from this perspective means dwelling upon all processes actually taking place in the body, including the shape that the mind takes because of the influences of the social relationships that it creates. Therefore, the question "What is the mind?" has to be tackled by answering the question "How does the mind work?", therefore starting from the identification of all mechanisms that make up the mind and allow it to work.

6.2   Fast and Automatic Versus Slow and Reflective

Starting from the 1970s, several authors (cfr. Evans 2007; Johnson-Laird 2006; Kahneman 2003; Kahneman and Frederick 2002; Sloman 1996; Stanovich 2004) developed—each with his own perspective—different versions of a theory that could be classified under the umbrella of the "dual process theory", which postulated a clear distinction between the cognitive processes that our mind performs swiftly, automatically, and unconsciously, and those that on the other hand require a slow, deliberative, and conscious processing.

Historically, the triggering observation of "dual process" theories was that a series of psychological tests revealed a repeated clash between logical and non-logical processes in the choice behaviour of test subjects tasked with a series of deductive reasoning exercises (Evans 1977). The famous Wason's test represents the most known example of this kind of event (1966).

In its original form, this test required the participants to select the cards that provided the right answer to an indicative or descriptive rule (which expressed a relationship between two states) such as "if p then q" from a deck. The participants were instructed to follow the rules of the test and were provided with four two-side printed cards showing, on one side, the information about the presence/absence of the first item (p or non—p), and on the other the information about the presence/absence of a consequence (q or non—q). The cards were then placed on a table with one face up (showing A, D, 3, 7). Then, the subjects were asked to identify the cards that they needed to turn in order to verify whether the rule "If one card shows A on one side, then 3 is printed on the other side" was true or false.

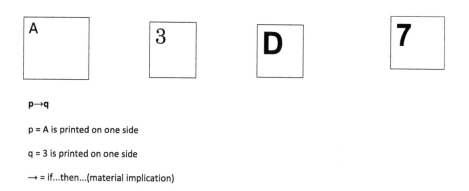

p→q

p = A is printed on one side

q = 3 is printed on one side

→ = if...then...(material implication)

The task at hand required the subjects to understand that the rule is false when A is accompanied by a number other than 3 on the other side. Hence, they had to turn the cards (A and 7), to verify whether this was not the case, forgetting about D and 3. Conversely, Wason's test outcomes showed that subjects tended to look for a corroboration of the rule rather than working the other way around, following an inefficient strategy. According to the results, only 10% of the participants provided the right answer (Evans 1989; Evans et al. 1993; Griggs and Cox 1983).

Despite the criticism put forward against different versions of this test over the last 40 years (Evans 2007; Oaksford and Chater 2007; Singley and Anderson 1989), it still represents the standard test for those wishing to study the limitations of subjects in providing a logically correct answer in the case of *modus tollens* (if p then q; non q then non p, cfr. Evans et al 1993). An interpretation of the apparently mixed outcomes of Wason's selection test was provided by Wason himself and Evans (1975) only few years after the first test. The authors claimed that the participants initially selected the cards following a primitive "matching bias" approach, due to the focus that some subjects placed on some elements openly mentioned in the conditioned statement. The authors came to the conclusion that the matching bias was an unconscious process that influenced the subjects' first reaction, while the motivations given to justify the choice were the product of a *post hoc* rationalisation process. This fact led the authors to use for the first time the terms of "type 1" and "type 2" processing to refer respectively to the first unconscious process and the second conscious and rational process.

Even though these remarks paved the way to the "dual process" theories, the discussion stalled for more than 20 years. Only after the publishing of a new book by Evans and David Over (1996) and an essay by Steven Sloman (1996) that referred to the article of Wason and Evans, was it possible to restart the conversation about this subject.

From a chronological perspective, the "dual process" theory by Evans and Over is the oldest one and, probably, also the one that best justifies the empirical data collected by the studies of psychology of reasoning. The authors start from the assumption that there is a distinction between an efficient answer-oriented process and the rationalisation process, which is typical when the belief-bias plays a substantial role. This distinction was firstly introduced in the article written by Wason and Evans in 1975. As it is well known, belief-bias become apparent when, in reasoning acts (e.g. syllogistic reasoning), old beliefs push subjects to provide answers that go against logics (examples of this effect are provided in the article written by Wason and Evans).

The belief bias is therefore, according to the authors, the proof that there are two types of reasoning processes: one that is influenced by the beliefs acquired in the past, the other evaluating the validity of logical arguments. The first system is defined "heuristic system" by the authors because despite its efficiency in providing good answers, it is still based on imperfect processes that often lead to mistakes. The second system is called "analytical system" and, contrary to the first one, it is able to analyse the elements of a statement and a decision by following logical rules, providing mostly right answers from a normative perspective.

Evans' approach seems therefore to have changed compared to the article written 20 years earlier with Wason. In his previous article, Evans stated that the second system was the only mechanism responsible for rationalisation, while in the book written with Over, the second system is merely a cognitive mechanism that ensures the conformity with the rules of logics. A similar change is noticeable also in the first system, which went from being a process that generally led to systemic errors, to an efficient mechanism that drives the majority of our actions, meeting the non-secondary objective of reducing the cognitive burden of some tasks.

It is also true that recognising the efficiency of heuristic mechanisms does not hinder the authors from stating their preference for the analytical processes that can provide normative correct answers. Nevertheless, despite this clear preference, the authors do recognise that the heuristic system also has its logics, introducing a distinction between the two types of rationality (rationality 1 vs. rationality 2) that are linked to each system. The authors do not mention the possible correlations between the two systems. Evans and Over limit themselves to criticising an hypothetical sequential model where heuristic techniques always precede analytical processes, admitting that there is also a "return possibility" between the two processes and hence a potential interaction between them (this aspect was thoroughly addressed and explained by Evans in a recent version of his book, cfr. Evans 2007).

Also in Sloman's theory, the references to type 1 and type 2 systems identified by Wason and Evans disappear. In this case, the author decided to use the terms "associative system" and "rule-based system" to distinguish between the two different cognitive processes. Nevertheless, he stated that:

> Both systems seem to try, at least some of the time, to generate a response. The rule-based system can suppress the response of the associative system in the sense that it can overrule it. However, the associative system always has its opinion heard and, because of its speed and efficiency, often precedes and thus neutralizes the rule-based response. (Sloman 2002, p. 391)

Sloman's work had a strong impact on the academic world and it supported the popularisation of the dual process approach. Nevertheless, Sloman focused on a field of study much more limited than the one dealing with dual system approaches, since the author was only interested in the analysis of reasoning and judgment and did not wish to address important topics such as the link between the architecture underlying both systems and the evolutionary debate (the same constraint can be found in other studies, such as those by Kahneman and Frederich (2002), who started from the analysis of some lab experiments that postulated that the heuristic system was inferior to the analytical system, because the former one is often associated to errors while the latter provides normative correct answers). Contrary to Sloman, the first researchers that did address the relationship between the dual process theory and evolution were Keith Stanovich and his partner Rich West (Stanovich and West 1997, 1998, 2003), who contended that evolution did not necessarily provide adaptation advantages, because the current environment in which humans live is completely different to the one in which evolution took place. This is the reason why the procedures related to the first system can often lead to a cognitive bias, particularly when focussing on abstract and decontextualised forms of reasoning. Nevertheless, contrary to animals, humans have a second system that allows individuals to pursue their objectives, even when these are not in line with those set by evolution. To specify the two systems, Stanovich coined the terms System 1 and System 2, which are the terms most used in academic literature. System 1 is described as a universal form of cognition, shared by humans and animals, with not only one system operating, but rather a set of subsystems (called TASS, The Autonomous Set of Systems), which work with a certain degree of autonomy. Because of the fact that System 1 is made of different subsystems, it can manage a high amount of information by processing it in parallel, an ability lacking to System 2.

The newest system (System 2) is from an evolutionary perspective an exclusively human system and, according to Stanovich, it is linked to intelligence. Despite its slow functioning, this system allows for abstract and hypothetical thinking, something that System 1 cannot do. Nevertheless, according to the author, despite System 1 is present both in humans and animals, it works differently when System 2 is present. Stanovich (2004) postulated that the knowledge that can be inferred through reasoning thanks to System 2 can also be "installed" in the implicit-automatic processing mechanisms of System 1 through the repetition of actions.

According to Evans and Frankish (2009), these three theories were developed independently. The only common element that they share is that they all postulate the existence of two different processes for performing specific tasks, based on two different procedures that lead to different and sometimes even contradicting results. As we have seen, the authors coined different names to label these two types of thinking. Others put forward a distinction between the two types of processes without postulating their underlying cognitive systems. The differences of all these descriptions are summed up in the tables below, taken from an article published by Evans in 2008 (Fig. 6.1).

The two tables developed by Evans (2008) show that not all authors agree on the features of the two systems. Nevertheless, the features postulated by all theories—despite focusing on different aspects of the same phenomenon—aim at the same

References	System 1	System 2
Fodor (1983, 2001)	Input modules	Higher cognition
Schneider & Schiffrin (1977)	Automatic	Controlled
Epstein (1994), Epstein & Pacini (1999)	Experiential	Rational
Chaiken (1980), Chen & Chaiken (1999)	Heuristic	Systematic
Reber (1993), Evans & Over (1996)	Implicit/tacit	Explicit
Evans (1989, 2006)	Heuristic	Analytic
Sloman (1996), Smith & DeCoster (2000)	Associative	Rule based
Hammond (1996)	Intuitive	Analytic
Stanovich (1999, 2004)	System 1 (TASS)	System 2 (Analytic)
Nisbett et al. (2001)	Holistic	Analytic
Wilson (2002)	Adaptive unconscious	Conscious
Lieberman (2003)	Reflexive	Reflective
Toates (2006)	Stimulus bound	Higher order
Strack & Deustch (2004)	Impulsive	Reflective

System 1	System 2
Cluster 1 (Consciousness)	
Unconscious (preconscious)	Conscious
Implicit	Explicit
Automatic	Controlled
Low effort	High effort
Rapid	Slow
High capacity	Low capacity
Default process	Inhibitory
Holistic, perceptual	Analytic, reflective
Cluster 2 (Evolution)	
Evolutionarily old	Evolutionarily recent
Evolutionary rationality	Individual rationality
Shared with animals	Uniquely human
Nonverbal	Linked to language
Modular cognition	Fluid intelligence
Cluster 3 (Functional characteristics)	
Associative	Rule based
Domain specific	Domain general
Contextualized	Abstract
Pragmatic	Logical
Parallel	Sequential
Stereotypical	Egalitarian
Cluster 4 (Individual differences)	
Universal	Heritable
Independent of general intelligence	Linked to general intelligence
Independent of working memory	Limited by working memory capacity

Fig. 6.1 Characteristics and features attributed to System 1 and System 2 by different authors (Evans 2008)

purpose, i.e. emphasising the difference existing between the two procedures that we have at our disposal to discover ourselves and the world surrounding us.

Going into details, despite the fact that the idea that System 2 is based on rules is in line with the proposals of the majority of dual process experts, the same cannot be said about the fact that Sloman claims that System 1 is associative. According to Evans, this is particularly true for the theories that oppose heuristic system and systemic (or analytical) processing, describing it as something different to associative processing.

Generally speaking, the recurrent themes tackled by academics deal with the fact that the processes of System 1 are concrete, contextualised, or related to a specific domain, while on the contrary the processes of System 2 are abstract, decontextualised, or related to a general domain.

Yet, defining System 2 as abstract and decontextualised could sound as a paradox when these features are considered in relationship to the other features usually attributed to this system, such as the fact that it is slow, sequential, explicit, and rule-based. These features cannot be limited exclusively to abstract forms of reasoning (Sloman 2002; Verschueren et al 2005). It would probably be fairer to admit that, despite the fact that abstract reasoning requires the use of System 2, concrete contexts do not hinder its implementation. Furthermore, without considering the personal favourite theory among those outlined, it is clear that the concept of System 2 is much wider than the one of logic reasoning as inhibitor of pragmatic influences (due to System 1) and the ability to perform hypothetical thoughts through assumptions and mental simulations (this is probably the reason why the majority of dual process experts tend to use general terms such as "analytical" or "systematic" when describing the second system).

Similar problems also arise with the definition of System 1 as a simple mental process that works automatically and without placing a burden on the working memory. Indeed, there are different types of this kind of implicit processes (which work for perception, attention, language processing, and so on).

6.3 Parallel-Competitive Versus Default-Interventionist

Despite the fact that dual process theories can count on strong empirical evidence coming from different disciplines of psychology, the concept of two cognition systems as defined by the idea of System 1 and 2 in the current literature is probably wrong, even though it is the concept with the strongest apparent appeal and a point of reference for all experts, as highlighted by Evans (2012). For example, it is probably wrong to consider System 1 a single and old system shared with other animals.

Similarly, it is probably wrong to consider System 2 the conscious mind, where all slow and sequential thinking takes place. Going into details, Evans (2012) identified five fallacies related to dual process theories:

(1) All dual process theories are essentially the same;
(2) There are just two systems underlying Type 1 and Type 2 processing;
(3) Type 1 processes are responsible for cognitive biases; type 2 processes are responsible for normatively correct responding;
(4) Type 1 processing is contextualised, while type 2 processing is abstract;
(5) Fast processing indicates the use of type 1 rather than type 2 processes.

Generally speaking, going beyond the constraints of these vague theoretical terms, an important development for dual process and dual system theories was represented by the input offered by neuroscience over the last few years. From this perspective, the identification of reflexive (System 1) and reflective (System 2) cognitive processing represented a very important step. These two types of processing were identified by cognitive neuroscientist Lieberman (2003), who attributed them to two neurological systems called System X and System C. Lieberman described the former as a system composed by the amygdale, basal ganglia, and the lateral temporal cortex. These cerebral regions were known to be involved in conditioning and associative learning and the author refers to them to explain the social cognitive processes usually described as automatic or implicit. Secondly, Lieberman refers to System C as responsible for explicit learning and the inhibitory operating control, contending that this system is composed by the anterior cingulate cortex, the prefrontal cortex, and the medial-temporal lobe (including the hippocampus).

The idea that our brain is composed of different parts, each devoted to different tasks or at least sequentially involved in the processing of an answer, does not shed light on whether and how the different cerebral regions interact with each other. According to Evans (2007), the academic literature provides examples of two modalities of interaction, which led to at least two different models. The first one is called "parallel-competitive" and postulates that System 1 and System 2 provide two different types of knowledge (implicit and explicit) competing with each other when providing an answer.

The second model is called "default-interventionist" and postulates instead that System 1 is always active, hence offering a gross and cost-effective processing tool to provide a preliminary answer, while System 2 is purposefully activated on the basis of the information provided by System 1. According to Evans, three factors determine the involvement of System 2 and the possibility of changing the default answer provided by System 1: education, general intelligence, time available. The stronger the focus attributed by education on the necessity of finding a logical answer, the bigger the intelligence of subjects and the higher the time needed to provide an answer. In this case, the analytical system may (and shall) get involved in order to change the answer given by the heuristic system. In addition to Evans' theory, Stanovich and Kahneman's theories also fall within the category of "default-interventionist" theories, while on the other hand Sloman's theory belongs to the "parallel-competitive" family.

To begin with, Evans and Stanovich (2013) defined parallel-competitive and default-interventionist types of designs as follows:

S. A. Sloman (1996; Barbey and Sloman, 2007)... proposed an architecture that has a parallel-competitive form. That is, Sloman's theories and others of similar structure (e.g.,

Smith and De Coster 2000) assume that Type 1 and 2 processing proceed in parallel, each having their say with conflict resolved if necessary. In contrast, our own theories (in common with others, most notably that of Kahneman and Frederick 2002; see also, Kahneman, 2011) are default-interventionist in structure (a term originally coined by Evans 2007). Default-interventionist theories assume that fast Type 1 processing generates intuitive default responses on which subsequent reflective Type 2 processing may or may not intervene. (Evans and Stanovich 2013, p. 227)

In the case of "Default-interventionist" models, System 1 works much faster and gives intuitive and environmentally sound answers that guide behaviour. Yet, in some instances, the answers are not fit for purpose and System 2 can replace them with more reflexive answers. A graphical representation of a stylised and simplified default-interventionist model is provided by the following figure (Fig. 6.2):

On the contrary, in "Parallel-competitive" models, System 1 and System 2 work continuously, they do not stop, and they also work simultaneously rather than sequentially, competing for the answer. A simple graphical representation of the features of a "parallel-competitive" model can be as follows (Fig. 6.3):

There is another theory not previously mentioned that belongs to the family of "parallel-competitive" models and could be useful in our reasoning; the theory developed by social psychologist Seymour Epstein (Epstein 1994; Epstein et al 1996; Epstein and Pacini 1999), which is quite peculiar in describing the functioning of System 1 and System 2 by introducing the idea of two competing processing styles.

Epstein is one of the few (or even the only one) that highlighted the experience-based nature of System 1 (putting it in opposition to a module-based approach), which might reflect an implicit learning process stored in parallel functional neuronal networks (Dijksterhuis and Smith 2005; Smith and De Coster 2000). His theory is known under the acronym "CEST" (Cognitive Experimental Self Theory) and assumes the existence of two systems, a cognitive-empirical one and a rational one. The former is described as the oldest one from an evolutionary point of view, with clear reference to animal cognition, while the latter is the newest one and it is typically human. Furthermore, each system has access to different types of knowledge. After assessing different cognitive studies and after carrying out several experimental (and psychometric) studies, Epstein came to the

Fig. 6.2 Default-interventionist model

Fig. 6.3 Parallel-competitive model

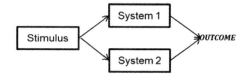

conclusion that the deliberative and analytical conscious system—that can reasonably be called rational system—works in opposition to the experience-based system, which is limited to the processing of non-verbal information (such as the emotional exponential processing caused by verbal inputs). Epstein contends that, during their evolution, higher organisms replaced instincts with a cognitive system that guides their behaviour on the basis of past experiences and lessons learnt. This system works differently to the one that has been developed much later and uses symbols and logical inferences to solve problems.

Obviously, we should not fall into the trap (as Epstein seems to do) of mixing up the theories postulating two thinking "modalities" or styles with theories postulating two different "types" of processes based on different cognitive systems. Only the latter can be considered dual process theories, while the others are quite widespread in psychology, particularly social psychology. "Types" are indeed related to the cognitive architecture, while "modalities" can be related to differences in personality, culture, and so on.

This kind of confusion and other forms of easy criticism could be avoided by looking for some "key" features that loosely define the two systems without going too much into detail and without being too vague or inconsistent. These key features can be identified following a hint given by the philosopher of the mind Pietro Perconti, who tried to define common sense (Perconti 2015). According to him, there are salient traits that play a role in distinguishing two levels of common sense; a deep level and a superficial one. These traits are: representation format, reference, resistance to change, and universality. The author presents them in the following format (Fig. 6.4):

	DEEP LEVEL	SUPERFICIAL LEVEL
REPRESENTATION FORMAT	Procedural schemas Metaphorical frames based on imaginative bodily representations Basic adaptive behaviour	Propositions and judgments
REFERENCE	Efficacy	Truthfulness and justifications
RESISTENCE TO CHANGE	Unchangeable through propositional knowledge	Modifiable
UNIVERSALITY	Universal and based on human biology	Sensitive to cultural variations

Fig. 6.4 Salient traits of the deep and superficial level of common sense (Perconti 2015)

According to Perconti, the deep level (which we might call System 1) hosts the kind of knowledge that allows subjects (either humans or animals) to fit to their environment. This level is naturally inclined to action. On the contrary, the superficial level hosts events recorded in the consciousness and presenting features associated to operating processes and higher conscious control.

Thanks to this distinction, Perconti is able to better match the deep level with all automatic cerebral processes (hence without excluding numerous different types of implicit processes related to perception, attention, language processing, and so on). This type of world-related knowledge cannot be recalled as explicit knowledge, but it can directly affect behaviour. Furthermore, it does not exclude habits and automatic behavioural models that in the past had required a conscious effort (type 2) but through practice and experience have become patterns that are deeply rooted in the deep level.

According to Perconti, the deep level can indeed be seen as the kind of non-conceptual knowledge that, contrary to the content of propositions and judgments, is made up by procedural schemas. The concept of schema used by the author is taken from Kant and it represents the "rule" used to create mental representations. Perconti highlights that "A schema does not replace something else, but it represents a procedure to produce the representations needed" (Perconti 2015, pp. 60–61).

The famous passage of Kant's Critique of Pure Reason mentioned by Perconti is the following:

> Thus, if I place five points in a row…. this is an image of the number five. On the contrary, if I only think a number in general, which could be five or a hundred, this thinking is more the representation of a method of representing a multitude (e.g., a thousand) in accordance with a certain concept than the image itself, which in this case I could survey and compare with the concept only with difficulty. Now this representation of a general procedure of the imagination for providing a concept with its image is what I call the schema for this concept. (Kant 1998, p. 273)

Kantian schemas are not contents of thought, but rather procedures to create contents and articulate them in classes. To clarify even more the concept of schema, Perconti draws an analogy with arrow signs: arrow signs do not convey information on the kind of item to which they point, they merely indicate the path that you need to follow to find it. Schemas are therefore a structured set of "instructions" to meet a target.

Contrary to schemas, pictures are proper mental representations, they equal to the act of thinking about facts, true or false representations. In his famous proposition number 2.1 in the Tractatus, Wittgenstein states "We picture facts to ourselves", while number 2.11 reads "A picture presents a situation in logical space". Summing up, Wittgenstein will never stop claiming that thinking and talking means producing pictures (Wittgenstein 1961).

Another author that seems to truly share Wittgenstein's line of thought about the role of pictures and icons is Peirce. When talking about mathematical knowledge (exact mathematical knowledge, of course), he states that this kind of knowledge is not a simple contentless formal game filled with empty formulae that must be interpreted, but it is rather the product of icons of possible reality status. In his words:

Thus, an algebraic formula is an icon, rendered such by the rules of commutation, association, and distribution of the symbols. It may seem at first glance that it is an arbitrary classification to call an algebraic expression an icon; that it might as well, or better, be regarded as a compound conventional sign. But it is not so. For a great distinguishing property of the icon is that by the direct observation of it other truths concerning its object can be discovered than those which suffice to determine its construction. Thus, by means of two photographs a map can be drawn, etc. Given a conventional or other general sign of an object, to deduce any other truth than that which it explicitly signifies, it is necessary, in all cases, to replace that sign by an icon. This capacity of revealing unexpected truth is precisely that wherein the utility of algebraical formulae consists, so that the iconic character is the prevailing one. (Peirce 1931, 2.279)

Further on:

In fact, every algebraical equation is an icon, in so far as it *exhibits,* by means of the algebraical signs (which are not themselves icons), the relations of the quantities concerned. (Peirce 1931, 2.282)

Icons are therefore an interesting phenomenon in the study of the relationships between language and reasoning, because the mere possibility of identifying a bond between signifier and meaning supports the idea that language reflects reasoning, which in turn reflects reality. In addition to this, icons allow us to step away from the subjectivity of linguistic signs, i.e. the fact that the association of signifier and meaning is not justified, but rather conventional. In opposition to this idea, Peirce highlights that, despite being overlooked and considered of secondary importance, icons play an influential role in the structuring of linguistic codes.

Nevertheless, despite the importance of icons as pillars of linguistic concepts, at a mathematical level—even when concepts are represented as icons—they cannot be fully considered mathematical concepts. In mathematics, icons are abstractions and, as such, they are not entities made of matter or energy. In a nutshell: many direct experiences can lead to a mathematical concept, but the mathematical concept is not any of these experiences. Let's take this example as a case in point: "Donald is on the fourth step of a set of stairs. If he climbs three steps up, to which step will he get?"

Children will have no particular problem in analysing this problem and they will soon find out how to solve it. If they have a set of stairs nearby, they could solve it with a direct action, moving up the stairs and checking the final position of the subject after following the instructions. Nevertheless, this only applies to examples such as the one outlined above, in which the problem is a procedure rather than a true mathematical concept.

As a matter of fact, children start struggling when they deal with mathematical problems that cannot be solved with a direct action, cases in which the problems are expressed and coded by symbols, decoupled from first-hand reality, such as:

$$4 + 3 = 7$$

Symbolic mathematical writing represents the end point, i.e. the product of an awareness-building and learning process, not the starting point to discover and master the mathematical concept of "addition".

The solution that can be reached by climbing the stairs allows children to access the signifiers that help in solving the problem without wasting time and energy performing other actions. In these cases, the direct process makes the problem easily solvable and manageable, leading to satisfying results both at understanding the concepts at hand and the ability to perform actions, because it creates a bond between languages that is more related to motory experience.

Nevertheless, these mathematical concepts are basically "ideal" and they are usually presented to children without referring to reality, but focussing on the possible relationships between objects or sets that are formally defined. The ideal character of mathematics implies that, particularly in schools, mathematical concepts are mainly built as wordy descriptions, in order to come to the most advanced form of mathematical concept that is represented by the "definition". From this point of view, mathematics is a language made up of technical terms and symbols that structure its discourse in a precise and severe manner. In order to "perform" mathematics, children must be introduced to this technical language, characterised by a severely codified formality that—contrary to natural language—is not learnt spontaneously while growing up. Therefore, it is not so unusual that children do not understand the utility and meaning of symbols and syntactical rules used to connect signifiers and meanings and end up hating mathematics.

The characters of "abstraction" and "idealisation" that define mathematics as a rigid (and even hated) discipline pushed several philosophers to see mathematical objects as absolutely certain and defined objects, since mathematical propositions (the product of a process that goes beyond reality) were considered as a set of truths, certain and undisputable truths. This assumption stems from the idea that while a physical hypothesis might be verified through an accurate laboratory test and its interpretation, mathematical truths are established through demonstrations based on axiomatic methods. This procedure should then provide mathematical propositions with the highest degree of reliability possible for human kind (Jaffe 1997). As a matter of fact, while in physics nothing is 100% certain, in mathematics —thanks to the axiomatic method—everything is certain for eternity (Jaffe 1997). These remarks on the concept of absolutely certain truths have been contested over the years (cfr. Hamming 1980; Leary 2000). Yet, the fact that mathematics does not have the absolute certain character of reliability that was postulated in the previous century, does not jeopardise the fact that it has an objective content (its content is indeed not made up of absolutely certain truths, but rather of plausible statements in line with current scientific knowledge, cfr. Cellucci 2017). As Solomon Feferman stated, mathematics is the paradigm of certain and well-understood knowledge, a coherent set of truths soundly bonded together, but not unchangeable (Feferman 1998).

It is useful to recall that our guiding principle is the one outlined by Perconti and shown in Fig. 6.4. Recognising mathematics as a paradigm of knowledge developed over time, but potentially changeable, makes it indeed a privileged domain of cultural evolution and, therefore, sensitive to time variations. Cultural evolution follows a pace that is totally different to the one of biological evolution. Even though humans are what they are because of biological evolution, they are like this also because cultural evolution provided them with cultural, social, and technological outcomes that define modern societies. By abiding by the limits established by biological evolution, cultural innovations (contrary to genetic mutations that are random, cfr. Gould 2002) often follow a well-defined direction and they are determined by the conscious intervention of individuals. This does not mean that they are always positive for the subjects or groups exploiting them and, indeed, natural selection sometimes "interferes" with the entire process and makes some noxious cultural traits disappear.

Starting from these assumptions, it is easy to point out that the mathematics outlined by Dehaene (and other cognitive scientists) has a much more limited scope, which merely involves the processes related to the estimation of numerosity, a term that refers to the sense of number and in particular to the sense of a size of sets, not of numbers. Dehaene's experiments only shed light on the existence of a natural, inborn, biologically founded form of mathematics shared by humans and animals, active also in children during their prelinguistic stage. According to this theory, natural selection and nature shaped animals and living creatures in such a way that they could perform actions and natural mathematical computation operations aimed at promoting their survival. In other words, the concept of mathematics expressed by Dehaene is similar to the concept of deep level of common sense described by Perconti, since the notion of numerosity refers only to the non-symbolic cardinality that allows animals (all animals) to adopt basic adaptive behaviours in a chaotic and irregular physical world. Besides, all knowledge, including mathematical knowledge, is part of a natural adaptation process to the surrounding environment. Yet, contrary to numerosity, in mathematics numbers are abstract entities represented by specific written and oral symbols that can be used to perform computational tasks.

As the ethologist John Bonner (1980) clarified, the difference between the two types of "information" can be noticed because of the fact that the cultural dimension depends from *possible* choices. Behaviour is cultural when (a) there is a number of possibilities at its basis, (b) it is the product of conscious choices, (c) these choices are shared by a specific group, (d) it is socially (rather than genetically) conveyed through imitation or proper procedures of teaching/learning. All these conditions require a low level of or almost zero genetic determination of the behaviour. The more behaviour is genetically determined, the less space there is for cultural information, and the other way around.

This will and possibility are not present in animals and, therefore, this difference creates a discontinuity. Obviously, it does not represent a biological discontinuity, but rather a discontinuity due to the fact that humans are the only creatures in nature that are subject to biological and cultural evolution.

To sum up, (system 2) mathematics can be considered as an artefact that was created in relation to the very fast pace of cultural development, a fact that clearly shows the "double nature" of humans (Tomasello 1999). This will be the topic tackled in the following paragraph.

6.4 "No Nature, One Nature, Two Natures"

Philosophers see mathematics as the queen of all scientific disciplines. It has specific features that make it unique. Chemistry, physics, and biology study reality from different points of view, while mathematics mainly deals with abstract entities (whatever abstract means) such as numbers and sets (and many more), setting it apart from the other fields of knowledge.

According to this dominant point of view, the reflection upon mathematics requires a specialised discipline (philosophy of mathematics) that follows a specific methodology (the unavoidable deductive logics—essence of demonstration tests—and axioms) and, as stated by Dummett, that represents "the easiest part of philosophy" from a certain point of view, because many general problems tackled by other study fields of philosophy can be found in the philosophy of mathematics in a purer or particularly simplified form. Therefore, Dummett asks "If you cannot solve these problems, what philosophical problems can you hope to solve?" (Dummett 1998).

At the basis of Dummett's optimism, there is the conviction that—contrary to what happens in other disciplines that infer their concepts from experience, observation, and scientific theories—mathematics does not follow the same procedure. It does not require inputs from experience, since it is an exclusive product of thought. In this way, Dummett postulates a "division of work" between philosophy and science, implicitly recalling the Diltheyan distinction between *Erklaren* and *Verstehen*, giving legitimacy to the dichotomy of knowledge procedures. According to Dummett, knowing the meaning of a mathematical expression means knowing how to prove or demonstrate the expression. On the contrary, an empiric expression such as "It is raining" is simply linked to the knowledge related to the ability of recognising situations where perception allows us to come to the conclusion that it is raining.

This point of view hence recognises two possibility domains that can lead to the knowledge of the world. On the one hand, the domain of Pure Reason, tasked with the control of the logical consistency and correctness of reasoning, while on the other hand, there is the domain of Facts, tasked with the experimental control of the truth of empirical statements.

Nevertheless, Western philosophy questioned the soundness of both knowledge domains, up to the point that today several philosophers explicitly contest the existence of Pure Reason (just think about logics experts such as Gödel, language philosophers such as Wittgensten, or science philosophers such as Poppers; all promoters of this line of thinking).

On the other hand, scientists recognise that Facts have simply proved that they are just facts that could be different or presented in other ways (Heisenberg and his uncertainty relations, which are linked to Bohr's complementary principle). Yet, between philosophy and science, science has been the one to suffer the most from this unbridled crisis of certain knowledge (crisis of reason, weak thought, groundless knowledge, invented reality: these are only some of the labels used to designate the recurrent symptoms of several philosophical theses).

As Gadamer stated, "By the Seventeenth Century, what we now call philosophy is in a different situation. The philosophy has become a need of legitimating towards school science as it had never happened before" (Gadamer 1976, p. 13). On his side, Wittgenstein contends that "The totality of true propositions is the whole of natural science (or all of the natural sciences)" (Wittgenstein 1961, 4.11). Since philosophy does not belong to the natural sciences family, it does not have anything to dwell upon.

The abandonment of the idea of a First Philosophy, foreign to scientific inquiry and superior to it, finds its most radical expression in Quine's philosophical thesis called "Naturalism":

> Naturalism: abandonment of the goal of a first philosophy. It sees natural science as an inquiry into reality, fallible and corrigible but not answerable to any supra-scientific tribunal, and not in need of any justification beyond observation and the hypothetical-deductive method. (Quine 1981, p. 72)

Quine's Naturalism (also known by the name of scientific Naturalism or radical Naturalism) hence implies a recognition of the fact that reality must be specified and described in the framework of science, not that of philosophy.

Following a simplistic approach, we can sum up Quine's Naturalism by saying that scientific theories are the only tool that can provide legit justification to the beliefs we build about reality. From this point of view, naturalist scientists should not dwell upon establishing epistemic rules, but rather limit their inquiries to a descriptive study of the causal process that takes humans from sensory stimulation to the establishment of beliefs.

Quine's Naturalism was followed by other forms of Naturalism, also in mathematics (cfr. Putnam 1971; Armstrong 1997; Weir 2005; Burgess and Rosen 1997, 2005). All these theories take the name of Naturalism, despite the fact that all of them take strongly different perspectives as their starting point. Among the most prominent thinkers of mathematical Naturalism, the American philosopher Penelope Maddy represents an important point of reference. In her case, she decided to build her form of Naturalism on sound cognitive foundations.

Taking inspiration from Quine's idea (even though she did not stay completely loyal to his spirit), Maddy defended a Naturalist (and Platonist) approach to mathematics, despite changing her stance over time. Yet, because of the particular field on which her studies focussed, soon enough she found herself forced to account for the entities tackled by the philosophers of mathematics, which at first sight cannot be linked to the entities postulated by natural sciences (in her specific case, the abstract entities of mathematics).

This is the so called "placement problem", i.e. the need to explain which place these entities have in the natural world. Maddy (1990) provided an answer to this specific question by supporting the idea of a mathematical Platonism (called "set-theoretic realism") and using a so called "indispensability argument" to justify the realism of mathematical entities, i.e. an argument that contends that the objective existence of abstract entities is an integral part of the best explanation available to us. Nevertheless, justifying Platonism meant opening up to mathematical intuition skills, which were strongly criticised by gnosiology from a naturalistic point of view. Maddy tried to solve the issue by supporting the idea that mathematical intuition is not only similar to sensory intuition, as Godel said, but also represents a proper perception skill, i.e. the perception of medium size sets containing physical objects, which are established in the brain (Maddy 1990). In this instance, the strategy followed by Maddy is a "second naturalistic strategy", i.e. a reductionist strategy, according to which these properties are ontologically identical, or dominant, to acceptable scientific properties. In her words:

> We perceive sets of physical objects much as we perceive the objects themselves. Both abilities develop gradually as the course of child-hood experience interacts with evolutionarily conditioned brain structures. The neurophysiologic changes that constitute this development also produce a range of extremely general belief about these sorts of things, about the space-occupying character of physical objects, for example, and the combinability of sets. (Maddy 1989, p. 1140)

The following year, in her book *Realism in mathematics*, Maddy refers specifically to the discoveries recently made by neuroscience and experimental psychology, in order to shed some light on the analogy between the intuition of sets and the perception of objects.

By recalling the discoveries of neurophysiologist Donald Hebb (1980)—who proved that neurons do not only perform immediate perceptive actions but, on the contrary, also send mutual electric stimuli well after the end of a sensory stimulus, creating so called "neural assemblies" (groups of neurons connected)—Maddy considers that these neural assemblies are the neurophysiologic correspondent of her idea of "physical object". In other words, the author claimed that, in order to create neural assemblies that could perceive an object, e.g. a "triangle", a subjects needs neural assemblies that identify angles, which can create assemblies that can identify a triangle from a specific perspective, which can lead to the creation of a perspective that combines all previous perspectives, finally connecting these assemblies to one single object. Maddy is convinced that these observations fully corroborate the theories put forward by Piaget, which related to the creation of concepts by children, from a neurophysiologic point of view.

In the same book, she writes:

> This expectation is substantiated by the experiments of Jean Piaget and his colleagues. The child's ability to acquire perceptual beliefs about physical objects, as judged from behaviour, develops between the ages of one and eighteen months. At the beginning of this period, the child's world is a welter of isolated incidents. (Maddy 1990, p. 54)

The author is hence claiming that the same neurons that are triggered by the repeated perception of an object from different perspectives keep firing signals to the others, establishing a neural assembly that works as "object detector". Starting from these considerations, particularly in relation to Piaget's seriation and correspondence experiments, Maddy postulates a similar procedure for the creation of the concept of set. In her words:

In this way, even an extremely complicated set would have spatial-temporal location, as long as it has physical things in its transitive closure. And any number of different sets would be located in the same place; for example, the set of the set of three eggs and the set of two hands is located in the same place as the set of the set of two eggs and the set of the other egg and the two hands. (Maddy 1990, p. 59)

Later on, Maddy (2005, 2007, 2011) denied that the indispensability argument could justify the existence of mathematical objects, taking a step back from Quine's Naturalism and defining it too restrictive towards mathematics, since it did not considered purely intramathematics acceptability standards as scientifically admissible. In *Second Philosopher*, Maddy supports indeed the idea of a pragmatic method developed by the mathematician community, trusting their judgment and ability to control the establishment of the theories belonging to their field of study. In *Second Philosopher*, despite keeping the starting point of Quine's Naturalism (and therefore the superiority of empirical sciences on the other forms of knowledge), the author indeed takes some distance from Quine's orthodoxy, realising that mathematics is embedded in empirical sciences and that it is therefore better to take into account the "whole mathematical practice", including the parts that do not directly belong to pure contemporary mathematics (Maddy 2005).

Despite these notable changes in perspective, also in *Second Philosopher*, Maddy uses the term "science" to refer to "cognitive sciences", with the difference that Piaget's studies are replaced by the neuroscientific discoveries made by Dehaene and Spelke on numerical cognition. Maddy emphasises that neuroscientific studies have provided interesting new and detailed inputs to the naturalistic research on mathematical entities, and that these studies have provided outcomes that diverge to the arguments she outlined in *Realism in Mathematics*. Nevertheless, the interpretation given by Maddy of these experiments is quite odd, since she considers them supporting evidence to the role played by the set theory in mathematics. The result is that Maddy succeeds in highlighting some cognitive invariable factors that, according to her Platonist interpretation of set theory, correspond to basic properties of the objects tackled by this theory.

This is not enough to provide a plausible epistemology to mathematics, since the role played by set theory in mathematics depends on the whole theory. Furthermore, as Parsons (2007) highlighted, even assuming that it is admissible to come to a conclusion from the description of perceptive phenomena that would be at the basis of elementary numbers, there would still be the problem that the mathematical theories applied to this description and those used in psychology and neuroscience can both still be used without referring to set theory. Therefore:

It is just not plausible that the formulation in terms of set theory reflects the nature of things to the degree that Maddy's view presupposes. (Parsons 2007, p. 211)

In the light of Parsons' remarks, it seems that the problem lies in the fact that an empirical justification cannot be given to the passage from empirically founded elementary mathematical beliefs to the processes on which mathematical theories are based, despite this being of crucial importance for the establishment of mathematics. This is where Maddy's scientific Naturalism fails. Besides, following the strong belief in the good faith of the mathematician community, Maddy ends up contradicting the main thesis of scientific Naturalism, which establishes the admissibility of a thesis considering exclusively its scientific utility for natural sciences.

Scientific Naturalists see natural sciences as the model to which all other sciences and even philosophy should adapt to be legitimated in their knowledge-building activity. Despite its character of legitimate criteria as far as its inspiring principles are concerned, this approach shows strong limitations when dealing with mathematical concepts. Several authors (cfr. Field 1989) believe that mathematical concepts such as those of number, set, etc. are so disconnected from the field of natural science that they cannot find space in it, and the fact that they cannot be addressed by natural science should suggest deleting them from the philosophical-scientific dictionary and replacing them with terms and concepts that have a stronger presence on a material level.

Luckily, over the last few years, a much less radical form of Naturalism compared to the scientific and eliminativist one is taking shape, the so-called moderate (or liberal) Naturalism. This form of Naturalism share what we may call the "founding thesis" of Naturalism, i.e. the use of laws, explanation models, and entities that exist in nature and do not belong to the supernatural world (religious or mystic beliefs, demiurges, gods, etc.). In addition to this, liberal naturalists share with scientific naturalists the idea that natural science represents the default model to which all other sciences should abide by in order to be legitimated in their knowledge-building activity.

Even though both approaches believe in the superiority of natural science and the use of the experimental data that it provides, they do differ on the role that philosophy should play. In liberal Naturalism, the cornerstone is the compatibility (rather than continuity) of philosophy and science, with a strong anti-reductionist character, particularly in relation to the topic of normativity. The supporter of a liberal form of Naturalism believes that scientific inputs are of paramount importance to philosophy and that philosophical formulations should take into account the discoveries of natural science, nevertheless he does not accept the continuity thesis of Quine's scientific Naturalism, because he claims that philosophy differs from science in its method, object study, and study goals (De Caro and Macarthur 2010).

Only by taking into account this fact is it possible to overcome the clear division —unavoidable in scientific Naturalism—between phenomena belonging purely to the physical world and those referring to other fields of human experience, recovering concepts (such as normativity, number, intentionality, free will) that cannot be connected to the physical world and providing them with dignity by asserting their belonging to the natural world, without recurring to metaphysical explanations.

But how can the normative and causal levels be reconciled, considering the usual importance attributed to causality in natural sciences? John McDowell tried to reply to this question with a theory that fully embodies the compromise solution supported by liberal naturalists. According to McDowell, the specificity of humans lies in their "second nature" (De Caro and Macarthur 2010). Taking inspiration from the concept of "space of reasons" of Wildrif Sellars, McDowell claims that the best way to explain some human behavioural traits is to refer not only to the "causes" that govern body movements, but also (and in particular) to the "reasons" of an action. Great care should be paid to the concept of "reasons", which should not be considered abstract and independent entities from human experience, but rather one of its components (humans' "second nature").

In this case, McDowell's liberal Naturalism meets the first precondition (which could be defined ontological) that made Maddy's scientific Naturalism crumble, i.e. the incursion among possible explanations of all kinds of entities needed to explain a fact, without pre-established constraints. From this perspective, there are no problems in accepting the existence of mathematical entities (and the truthfulness or falsehood of their judgment), on the condition that these entities are needed to explain important aspects of our thought and do not represent supernatural entities that violate the laws of nature. In the specific case of mathematical entities, it is important to avoid a "representation mistake" and assert once and for all the truthfulness of what's real according to our symbolic and mathematical beliefs. To clarify more the issue of mathematical entities, two different notions of existence are needed, and liberal Naturalism does not struggle in providing them with a clear explanation. This double concept of existence draws a line between r-existence and l-existence. To use the previous terminology, r-existence is the l-existence in the ordinary sense of the term. L-existence means belonging to a specific interpretation domain. It represents an existence with a linguistic nature and the objects that l-exist have an identity thanks to some linguistic criteria. To establish whether something l-exists, it is necessary to identify the object in the logic space of the discourses in which it is mentioned (Perconti 2003, p. 10).

Hence, liberal Naturalism (as proposed by McDowell and all other researchers that took inspiration from this form of Naturalism) does not struggle in accepting conceptual analysis (the second precondition, which has a methodological nature) as a legitimate inquiry method when this is useful to explain specific phenomena, on the condition that this method is not incompatible with natural science studies, such as neuroscientific inquiries. If this is true, then, normativity is not incompatible with descriptive and causal inquiries, which logically means that it is compatible with them.

Claiming that humans, differently from physical systems and animals (because of the lack of language), enjoy a "second nature" means, in conclusion, accepting as good epistemic practices also those that are compatible with the cognitive modalities that humans do have (first nature). This remark is true for all products of cognition, and therefore also for mathematics. In conclusion, we can indeed claim that, during its evolution, mathematical knowledge was affected *by* the brain as much as *for* the brain. It was influenced by the brain because the history of mathematics is linked to more and more advanced cerebral invention capacities that allowed for the development of new enumeration systems and arithmetical procedures. It was influenced for the brain because only the inventions that were most suited to the cognitive and mnemonic capacities of humans and could increase the computation abilities of humanity were culturally transmitted to the following generation.

References

Armstrong, D. (1997). *A world of states of affairs*. Cambridge: Cambridge University Press.

Barbey, A. K., & Sloman, S. A. (2007). Base-rate respect: From ecological validity to dual processes. *Behavioral and Brain Sciences, 30,* 241–297.

Bonner, J. T. (1980). *The evolution of culture in animals*. Princeton: Princeton University Press.

Burgess, J. P., & Rosen, G. (1997). *A subject with no object. Strategies for nominalistic interpretation of mathematics*. Oxford-New York: Oxford University Press.

Burgess, J. P., & Rosen, G. (2005). Nominalism reconsidered. In S. Shapiro (Ed.), *Oxford handbook of philosophy of mathematics and logic* (pp. 515–536). Oxford-New York: Oxford University Press.

Cellucci, C. (2013). Philosophy of mathematics: Making a fresh start. *Studies in History and Philosophy of Science Part A, 44*(1), 32–42.

Cellucci, C. (2017). *Rethinking knowledge. The heuristic view*. Berlin: Springer.

De Caro, M., & Macarthur, D. (2010). *Naturalism and normativity*. New York: Columbia University Press.

Dehaene, S. (2011). *The number sense. How the mind creates mathematics*, Revised and Updated Edition. New York: Oxford University Press.

Dijksterhuis, A., & Smith, P. K. (2005). What do we do unconsciously? And how? *Journal of Consumer Psychology, 15*(3), 225–229.

Dummett, M. (1998). The philosophy of mathematics. In A. Grayling (Ed.), *Philosophy 2. Further through the subject*, 122–196. Oxford: Oxford University Press.

Epstein, S. (1994). Integration of the cognitive and psychodynamic unconscious. *American Psychologist, 49,* 709–724.

Epstein, S., & Pacini, R. (1999). Some basic issues regarding dual-process theories from the perspective of cognitive-experiential theory. In S. Chaiken & Y. Trope (Eds.), *Dual-process theories in social psychology*. New York: Guilford.

Epstein, S., Pacini, R., Denes-Raj, V., & Heier, H. (1996). Individual differences in intuitive-experiential and analytic-rational thinking styles. *Journal of Personality and Social Psychology, 71,* 390–405.

Evans, J. S. B. T. (1977). Toward a statistical theory of reasoning. *Quarterly Journal of Experimental Psychology, 29,* 297–306.

Evans, J. S. B. T. (1989). *Bias in human reasoning: Causes and consequences*. Brighton: Erlbaum.

Evans, J. S. B. T. (2007). On the resolution of conflict in dual-process theories of reasoning. *Thinking & Reasoning, 13,* 321–329.

Evans, J. S. B. T. (2008). Dual-processing accounts of reasoning, judgment, and social cognition. *Annual Review of Psychology, 59,* 255–278.

Evans, J. S. B. T. (2012). Questions and challenges for the new psychology of reasoning. *Thinking & Reasoning, 18,* 5–31.

Evans, J. S. B. T., & Frankish, K. (2009). *In two minds: Dual processes and beyond*. Oxford: Oxford University Press.

Evans, J. S. B. T., & Over, D. E. (1996). *Rationality and reasoning*. Hove, England: Psychology Press.

Evans, J. S. B. T., & Stanovich, K. E. (2013). Dual-process theories of higher cognition: Advancing the debate. *Perspectives on Psychological Science, 8,* 223–241.

Evans, J. S. B. T., Newstead, S. E., & Byrne, R. M. J. (1993). *Human reasoning: The psychology of deduction*. New York: Erlbaum.

Feferman, S. (1998). *In the light of logic*. Oxford: Oxford University Press.

Field, H. (1989). *Realism, mathematics and modality*. Oxford: Blackwell.

Gadamer, H. G. (1976). *Vernunft im Zeitalter der Wissenschaft*. Frankfurt: Suhrkamp.

Gould, S. J. (2002). *The structure of evolutionary theory*. Cambridge, MA: Harvard University Press.

Griggs, R. A., & Cox, J. R. (1983). The effects of problem content and negation on Wason's selection task. *Quarterly Journal ofExperimental Psychology, 35,* 519–533.

Hamming, R. W. (1980). The unreasonable effectiveness of mathematics. *The American Mathematical Monthly, 87,* 81–90.

Hebb, D. O. (1980). *Essay on mind*. Hillsdale, NJ: Lawrence Erlbaum Associates.

Jaffe, A. (1997). Proof and the evolution of mathematics. *Synthese, 111,* 133–146.

Johnson-Laird, P. N. (2006). *How we reason*. Oxford: Oxford University Press.

Kahneman, D. (2003). A perspective on judgment and choice: Mapping bounded rationality. *American Psychologist, 58*(9), 697–720.

Kahneman, D. (2011). *Thinking, fast and slow*. New York, NY: Farrar, Straus and Giroux.

Kahneman, D., & Frederick, S. (2002). Representativeness revisited: Attribute substitution in intuitive judgement. In T. Gilovich, D. Griffin, & D. Kahneman (Eds.), *Heuristics and biases: The psychology of intuitive judgment* (pp. 49–81). Cambridge, UK: Cambridge University Press.

Kant, I. (1998). *Critique of pure reason. The Cambridge edition of the works of immanuel Kant*. Cambridge: Cambridge University Press.

Leary, C. C. (2000). *A friendly introduction to mathematical logic*. Saddle River, NJ: Prentice Hall, Upper.

Lieberman, M. D. (2003). Reflective and reflexive judgment processes: A social cognitive neuroscience approach. In J. P.Forgas, K. R. Williams, W. von Hippel (Eds.), *Social judgments: Implicit and explicit processes*. New York: Cambridge University Press.

Lo Piparo, F. (2014). Nome e numero: una parentela. *Versus, 118,* 185–196.

Maddy, P. (1989). The roots of contemporary platonism. *The Journal of Symbolic Logic,* 1121–1144.

Maddy, P. (1990). *Realism in mathematics*. Oxford: Oxford University Press.

Maddy, P. (2005). Three forms of naturalism. In S. Shapiro (Ed.), *Oxford handbook of philosophy of mathematics and logic* (pp. 437–460). Oxford-New York: Oxford University Press.

Maddy, P. (2007). *Second philosophy*. Oxford-New York: Oxford University Press.

Maddy, P. (2011). *Defending the axioms*. Oxford-New York: Oxford University Press.

Oaksford, M., & Chater, N. (2007). *Bayesian rationality: The probabilistic approach to human reasoning*. Oxford: Oxford University Press.

Parsons, C. (2007). *Mathematical thought and its objects*. Cambridge: Cambridge University Press.

Peirce, C. P. (1931). *Collected papers*. Cambridge: Harvard University Press.

Perconti, P. (2003). *Leggere le menti*. Milano: Bruno Mondadori Editore.

Perconti, P. (2015). *La prova del budino. Il secondo comune e la nuova scienza della mente*. Milano: Mondadori.

Putnam, H. (1971). *Philosophy of logic*. New York: Harper & Row.

Quine, W. O. (1981). *Theories and things*. Cambridge, MA: Harward University Press.

Singley, M. K., & Anderson, J. R. (1989). *The transfer of cognitive skill*. Cambridge: Harvard University Press.

Sloman, S. A. (1996). The empirical case for two systems of reasoning. *Psychological Bulletin, 119*(1), 3–22.

Sloman, S. A. (2002). Two systems of reasoning. In T. Gilovich, D. Griffin, & D. Kahneman (Eds.), *Heuristics and biases: The psychology of intuitive judgment*, 379–398. Cambridge, UK: Cambridge University Press.

Smith, E. R., & De Coster, J. (2000). Dual-process models in social and cognitive psychology: Conceptual integration and links to underlying memory systems. *Personality and Social Psychology Review, 4*(2), 108–131.

Stanovich, K. E. (2004). *The robot's rebellion: Finding meaning the age of Darwin.* Chicago: Chicago University Press.

Stanovich, K. E., & West, R. F. (1997). Reasoning independently of prior belief and individual differ-ences in actively open-minded thinking. *Journal of Educational Psychology, 89*(2), 342–357.

Stanovich, K. E., & West, R. F. (1998). Cognitive ability and variation in selection task performance. *Thinking Reasoning, 4,* 193–230.

Stanovich, K. E., & West, R. F. (2003). Evolutionary versus instrumental goals: How evolutionary psychology misconceives human rationality. In D. E. Over (Ed.), *Evolution and the psychology of thinking* (pp. 171–230). Hove, UK: Psychol. Press.

Tomasello, M. (1999). *The Cultural origins of human cognition.* Harvard: Harvard University Press.

Verschueren, N., Schaeken, W., & d'Ydewalle, G. (2005). A dual-process specification of causal conditional reasoning. *Think & Reasoning, 11*(3), 239–278.

Wason, P. C. (1966). Reasoning. In B. M. Foss (Ed.), *New horizons in psychology: I* (pp. 106–137). Harmandsworth, England: Penguin.

Wason, P. C., & Evans, J. St. B. T. (1975). Dual processes in reasoning? *Cognition, 3,* 141–154.

Weir, A. (2005). Naturalism reconsidered. In S. Shapiro (Ed.), *Oxford handbook of philosophy of mathematics and logic* (pp. 460–482). Oxford-New York: Oxford University Press.

Wittgenstein, L. (1961). *Tractatus logico-philosophicus.* London: Routledge.

Printed in the United States
By Bookmasters